SpringerBriefs in Quantitative Finance

SpringerBriefs present concise summaries of cutting-edge research and practical applications across a wide spectrum of fields. Featuring compact volumes of 50 to 125 pages, the series covers a range of content from professional to academic. Briefs are characterized by fast, global electronic dissemination, standard publishing contracts, standardized manuscript preparation and formatting guidelines, and expedited production schedules.

Typical topics might include:

- A timely report of state-of-the art techniques
- A bridge between new research results, as published in journal articles, and a contextual literature review
- A snapshot of a hot or emerging topic
- An in-depth case study

SpringerBriefs in Quantitative Finance showcase topics of current relevance in the field of mathematical finance in a compact format. Published titles will feature both academic-inspired work and more practitioner-oriented material, with a special focus on the application of recent mathematical techniques to finance, including areas such as derivatives pricing and financial engineering, risk measures and risk allocation, risk management and portfolio optimization, computational methods, and statistical modelling of financial data.

More information about this series at http://www.springer.com/series/8784

Alexandre Antonov · Michael Konikov ·
Michael Spector

Modern SABR Analytics

Formulas and Insights for Quants, Former Physicists and Mathematicians

Springer

Alexandre Antonov
Standard Chartered Bank
London, UK

Michael Konikov
Numerix LLC
New York, NY, USA

Michael Spector
Numerix LLC
New York, NY, USA

ISSN 2192-7006 ISSN 2192-7014 (electronic)
SpringerBriefs in Quantitative Finance
ISBN 978-3-030-10655-3 ISBN 978-3-030-10656-0 (eBook)
https://doi.org/10.1007/978-3-030-10656-0

Library of Congress Control Number: 2019934783

Mathematics Subject Classification (2010): 60H30, 91B24

This Springer imprint is published by the registered company Springer Nature Switzerland AG
The registered company address is: Gewerbestrasse 11, 6330 Cham, Switzerland

To our loved ones/нашим любимым

Contents

Chapter 1
Introduction

1.1 Introduction

The SABR model introduced in Hagan et al. [39] is widely used by practitioners to capture skew and smile features observed in the interest rates implied volatilities. The underlying forward rate process F_t is assumed to follow the Constant Elasticity of Variance (CEV) evolution [20] with log-normal stochastic volatility (SV) v_t

$$\begin{aligned} \mathrm{d}F_t &= F_t^{\beta}\, v_t\, \mathrm{d}W_t^1 \\ \mathrm{d}v_t &= \gamma\, v_t\, \mathrm{d}W_t^2 \end{aligned}$$

with some correlation $\mathbb{E}\left[\mathrm{d}W_t^1\, \mathrm{d}W_t^2\right] = \rho \mathrm{d}t$, power β, $0 \le \beta \le 1$, and absorbing boundary condition at zero. Following Hagan et al. [39] we denote the initial SV as $\alpha = v_0$. The coefficient γ is called volatility of volatility. Since SV process v_t is not directly observable, we have four model parameters: α, β, ρ, and γ.

The primary usage of the SABR model is volatility interpolation. For example, we may have quotes for 1Y10Y swaptions of several strikes, each swaption being a 1Y option on a forward swap with 10Y tenor and some fixed rate (strike). We calibrate our SABR model to these quotes (given in price or volatility terms), and use the resulting model to compute values of any other 1Y10Y swaption of a given strike. We will discuss later the other applications of the model.

In the original article, Hagan et al. [39] came up with an approximation for European option price with strike K expressed through the implied Black volatility

$$\begin{aligned} \sigma_B(K) = {} & \frac{\alpha}{(F_0 K)^{(1-\beta)/2} \left\{1 + \frac{(1-\beta)^2}{24} \log^2 F_0/K + \frac{(1-\beta)^4}{1920} \log^4 F_0/K + \cdots \right\}} \left(\frac{z}{x(z)}\right) \\ & \times \left\{1 + \left[\frac{(1-\beta)^2}{24} \frac{\alpha^2}{(F_0 K)^{1-\beta}} + \frac{1}{4} \frac{\rho \beta \gamma \alpha}{(F_0 K)^{(1-\beta)/2}} + \frac{2 - 3\rho^2}{24} \gamma^2 \right] T + \cdots \right\}, \end{aligned}$$

A. Antonov et al., *Modern SABR Analytics*,
SpringerBriefs in Quantitative Finance,
https://doi.org/10.1007/978-3-030-10656-0_1

where

$$z = \frac{\gamma}{\alpha}(F_0 K)^{(1-\beta)/2} \log F_0/K$$

and

$$x(z) = \log\left\{\frac{\sqrt{1-2\rho z + z^2} + z - \rho}{1-\rho}\right\},$$

with F_0 being the initial rate value. The ATM volatility, $\sigma_B(F_0)$, also called backbone, can be written as

$$\sigma_B(F_0) = \frac{\alpha}{F_0^{1-\beta}}\left\{1 + \left[\frac{(1-\beta)^2}{24}\frac{\alpha^2}{F_0^{2-2\beta}} + \frac{1}{4}\frac{\rho\beta\gamma\alpha}{F_0^{1-\beta}} + \frac{2-3\rho^2}{24}\gamma^2\right]T + \cdots\right\}.$$

The implied volatility formula simplifies for strikes close to the ATM rate, $K \sim F_0$,

$$\sigma_B(K) = \frac{\alpha}{F_0^{1-\beta}}\left\{1 - \frac{1}{2}(1-\beta-\rho\lambda)\log K/F_0 + \frac{1}{12}\left[(1-\beta^2) + (2-3\rho^2)\lambda^2\right]\log^2 K/F_0 + \cdots\right\},$$

where coefficient

$$\lambda = \frac{\gamma}{\alpha}F_0^{1-\beta}$$

is the ratio of the volatility of volatility over the approximate ATM volatility $\alpha F_0^{\beta-1}$.

The methodology was based on small time expansion and was refined later by many authors, including Beresticky et al. [17], Henry-Labordere [43], Paulot [67], and others. We present below a simple and intuitive derivation, partially based on Andreasen-Huge [6], of an analog of Hagan et al. formula but, first, we explain why this model is so popular in the industry.

1.2 Wide Popularity of the SABR

Because the SABR model allows an analytic approximation for Black volatilities—we will address the quality of this approximation later—it is widely used for the swaption volatility *cube* interpolation and extrapolation. Such cube encapsulates swaption quotes information and is usually endowed with a way to produce quotes for swaptions of arbitrary expiry, swap tenor, and strike, hence, an arbitrage free interpolation and extrapolation is needed. Usually, there are standard liquid swaption maturities and swap tenors, and for each such expiry T and swap tenor L combination, a SABR model with parameters $\{\alpha, \beta, \gamma, \rho\}$ is fitted to quotes observed in the market. The result is surfaces (in (T, L) coordinates) of calibrated SABR parameters. If a swaption corresponding to a point (T, L) on the surface needs to be priced, each SABR parameter gets interpolated, and a SABR model with interpolated parameters is used to compute volatility at swaption strike K. Given a large number of the cube "nodes" the analytical solution of the SABR is quite desirable. Moreover, the model is capable of fitting the smile even when power β is fixed and only 3 parameters $\{\alpha, \gamma, \rho\}$ are calibrated. Even if the number of strikes is quite large (>10) the 3 parameters are often sufficient to accurately reproduce the market information. One plausible explanation may be the fact that most of the market players trade the smile with ... the SABR.

Another attractive feature of the model is a clear physical meaning of the parameters and their relations with different modes of the smile. The most important, being the ATM volatility, skew, and curvature, have a simple 0-th order in time expression via the SABR parameters

$$\begin{aligned}
\text{ATM volatility} \rightarrow \quad & \sigma_B(f, f) = \sigma_0 := \alpha f^{\beta-1}, \\
\text{skew} \rightarrow \quad & \left.\frac{\partial \log \sigma_B(f, K)}{\partial \log K}\right|_{K=f} = \frac{\sigma_0}{2}(\beta - 1 + \rho\lambda), \\
\text{curvature} \rightarrow \quad & \left.\frac{\partial^2 \log \sigma_B(f, K)}{\partial \log K^2}\right|_{K=f} = \frac{\sigma_0}{6}\left[(\beta-1)^2 + (2-3\rho^2)\lambda^2\right].
\end{aligned}$$

The skew has two parts: the β one and the ρ one. The first part gives a negative skew ($\beta \leq 1$) while the second one depends on the correlation sign: for negative correlations the second part is negative while for the positive correlation it is positive and can eventually dominate the negative β part. Such positive skews are sometimes observed for low rates.

The curvature is dominated by the positive part $(\beta - 1)^2$ but, at least theoretically, can be negative due to the second "volga" term $(2 - 3\rho^2)\lambda^2$.

Another important property of the SABR model is its "right" smile dynamics, i.e. dependence of the implied volatility smile $\sigma_B(K, F_0)$ on the forward rate while keeping all the model parameters fixed. The importance of its natural behavior is obvious for hedging purposes. This corresponds to an observation that the smile moves in the same direction as the forward.

Below we plot a normal implied volatility $\sigma_N(K, F_0)$ for the SABR with parameters $T = 1$, $\beta = 1/2$, $\rho = -30\%$, $\gamma = 70\%$, $\alpha = 5.77\%$ while the initial value F_0 moves from 3 to 3.25%. We observe a smile movement to the right with the increasing ATM volatility according to the approximate formula $\sigma \sim \alpha F_0^{\beta}$ (Fig. 1.1).

As noted in [39] this property was an important advantage w.r.t. the local volatility model which has the opposite smile dynamics—the smile moves in the direction opposite to that of the forward rate.

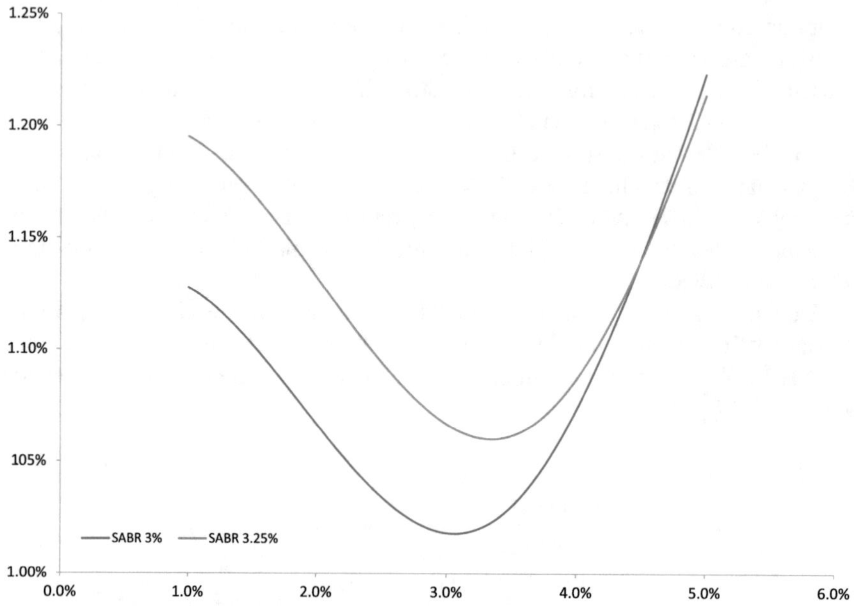

Fig. 1.1 Normal implied volatility movement

1.3 Simple Derivation

Consider a call option forward value

$$C(T, K) = \mathbb{E}\left[(F_T - K)^+\right].$$

Its time-derivative can be represented through the forward density

$$p(T, K) = \mathbb{E}\left[\delta(F_T - K)\right],$$

where $\delta(\cdot)$ is Dirac delta-function. Indeed, applying the Ito's lemma to a process $h_t = (F_t - K)^+$, we get

$$\mathrm{d}h_t = 1_{F_t > K}\, \mathrm{d}F_t + \frac{1}{2}\delta(F_t - K)\, \mathrm{d}F_t^2.$$

After averaging both sides of the equation, we obtain

$$\mathrm{d}\mathbb{E}\left[h_t\right] = \frac{1}{2}\mathbb{E}\left[\delta(F_t - K)\, v_t^2\, F_t^{2\beta}\right] \mathrm{d}t\,,$$

or

$$\frac{\partial C(T,K)}{\partial T} = \frac{1}{2} p(T,K)\, K^{2\beta} \mathbb{E}\left[v_T^2 \mid F_T = K\right] = \frac{1}{2}\frac{\partial^2 C(T,K)}{\partial K^2} K^{2\beta} \mathbb{E}\left[v_T^2 \mid F_T = K\right], \tag{1.1}$$

where we used the conditional expectation definition

$$\mathbb{E}\left[Y \mid X = x\right] = \frac{\mathbb{E}\left[Y\,\delta(X-x)\right]}{\mathbb{E}\left[\delta(X-x)\right]}.$$

The formula (1.1) is a special case of the Gyongy's lemma [38] for the effective local volatility

$$\sigma_{loc}^2(T,K) = \mathbb{E}\left[\sigma_{stoch}^2(T) \mid F_T = K\right] = \mathbb{E}\left[v_T^2 F_T^{2\beta} \mid F_T = K\right] = K^{2\beta}\mathbb{E}\left[v_T^2 \mid F_T = K\right].$$

Now, let us integrate it

$$C(T,K) = (F_0 - K)^+ + \frac{1}{2}K^{2\beta}\int_0^T \mathrm{d}t\; p(t,K)\,\mathbb{E}\left[v_t^2 \mid F_t = K\right] \tag{1.2}$$

and analyze for the small time. Indeed, the singular part is located in the density and has the form

$$p(T,K) \sim \mathrm{e}^{-\frac{s^2(F_0,K)}{2T}}$$

where $s(F_0, K)$ has a meaning of the "length" between the initial F_0 and the final points K (we will derive this formula in Chap. 4, but now we will obtain it using a heuristic approach). The conditional expectation of the stochastic volatility is obviously O(1) in small times and the time-integral is proportional in the main order in time to the density value at maturity, $p(T, K)$,

$$C(T,K) - (F_0 - K)^+ \sim p(T,K). \tag{1.3}$$

Now let us analyze the underlying density $p(T, K) = \mathbb{E}\left[\delta(F_T - K)\right]$. It will be easier to work with its conditional expectation

$$p(t, f, v; T, K) = \mathbb{E}\left[\delta(F_T - K) \mid F_t = f, v_t = v\right].$$

We will calculate the main asymptotics of the density using an intuitive approach from [6]. Indeed, imagine that we have a function $x(F, v)$ such that $x(K, v) = 0$ and a process $X_t = x(F_t, v_t)$ is Markovian with a known distribution. Then, we can relate the underlying densities

$$\begin{aligned}\mathbb{E}\left[\delta(F_T - K) \mid F_t = f, v_t = v\right] &\sim \mathbb{E}\left[\delta(x(F_T, v_T)) \mid x(F_t, v_t) = x(f,v)\right]\\ &= \mathbb{E}\left[\delta(X_T) \mid X_t = x(f,v)\right]\end{aligned}$$

in the main order in short maturities. Indeed, $\delta(F_T - K)$ is proportional to $\delta(x(F_T, v_T))$ modulo a *regular* derivative $\partial x(K, v)/\partial K$. Thus, this approximation does not change the main asymptotics after the averaging because the singularity remains the same.

In general, such process x is required to have a unit diffusion term, $\mathrm{d}x = \cdots \mathrm{d}t + \mathrm{d}W$ such that its small time distribution is Gaussian. This leads to a simple short maturity option time-value

$$O(T, K) = C(T, K) - (F_0 - K)^+ \sim p(T, K) \sim \mathrm{e}^{-\frac{x^2(F_0, v_0)}{2T}}. \tag{1.4}$$

Recall that the drift does not affect the main order of the small time asymptotics due to the scale difference, $\mathrm{d}W_t \sim \mathrm{d}t^{\frac{1}{2}} \gg \mathrm{d}t$.

As the first example, start with the local volatility model

$$\mathrm{d}F_t = \sigma(F_t)\,\mathrm{d}W_t.$$

As far as we can ignore the drift in our small-time limit we can operate with the SDE's as with the ordinary integrals to equate

$$\int \frac{\mathrm{d}F_t}{\sigma(F_t)} \simeq W_t + \text{const},$$

which gives the desired function

$$x(f) = -\int_f^K \frac{\mathrm{d}F}{\sigma(F)}, \tag{1.5}$$

which equals to zero for $f = K$. This formula is known as Lamperti transformation.

The corresponding process $X_t = x(F_t)$ has a unit diffusion coefficient,

$$\mathrm{d}X_t = \cdots \mathrm{d}t + \frac{\mathrm{d}F_t}{\sigma(F_t)} = \cdots \mathrm{d}t + \mathrm{d}W_t.$$

Thus,

$$p(T, K) \equiv p(0, F_0; T, K) \sim \mathbb{E}\left[\delta(x(F_T))\right] \sim \mathrm{e}^{-\frac{1}{2T}\left(\int_{F_0}^K \frac{\mathrm{d}F}{\sigma(F)}\right)^2}. \tag{1.6}$$

Now we are ready to proceed with the SABR process. First, inspired by the local volatility case, consider a process $Y_t = y(F_t, v_t)$ for

$$y(f, v) = -\frac{1}{v}\int_f^K \frac{\mathrm{d}F}{F^\beta} = \frac{f^{1-\beta} - K^{1-\beta}}{v\,(1-\beta)},$$

such that the SDE reads

$$\mathrm{d}Y_t = \cdots \mathrm{d}t + \frac{\mathrm{d}F_t}{v_t F_t^\beta} - Y_t \frac{\mathrm{d}v_t}{v_t} = \cdots \mathrm{d}t + \mathrm{d}W_t^1 - Y_t \gamma \mathrm{d}W_t^2 = \cdots \mathrm{d}t + \sqrt{1 - 2\rho\gamma Y_t + \gamma^2 Y_t^2}\,\mathrm{d}B_t,$$

where B_t is some Brownian motion. Sometimes, where needed, we will explicitly restore the strike argument separated with a vertical line

$$y(f, v \,|\, K) = \frac{f^{1-\beta} - K^{1-\beta}}{v\,(1-\beta)}\,.$$

Again, using the local volatility logic we come up with the desired function

$$\begin{aligned} x(f, v) &= \int_0^{y(f,v)} \frac{\mathrm{d}y}{\sqrt{1 - 2\rho\gamma y + \gamma^2 y^2}} \qquad (1.7)\\ &= \gamma^{-1} \log\left(\frac{\sqrt{1 - 2\rho\gamma y(f,v) + \gamma^2 y^2(f,v)} + \gamma y(f,v) - \rho}{1-\rho}\right). \end{aligned}$$

Thus, the option time-value asymptotics for small maturities is given by (1.4) for the process x defined above (1.7). It is important to notice that the option price calculated directly by the formula (1.4) has quite poor accuracy. One can dramatically improve it by a procedure called "mapping", see Chap. 4 for details. It consists of the following steps:

- Choose a calculation model with a known exact solution, e.g. Black model: $\mathrm{d}F = F\sigma_B\,\mathrm{d}W$ with the function $x_B(f) = \sigma_B^{-1} \log f/K$.
- Match the calculation model main small time-asymptotics for a given strike with the initial model ones

$$O_{SABR}(T, K) \sim O_B(T, K) \;\; \Rightarrow \;\; x_{\text{SABR}}(F_0, v_0) = x_B(F_0)$$

to find the resulting effective calculation model parameters—the Black volatility

$$\sigma_B = \frac{\log F_0/K}{x_{\text{SABR}}(F_0, v_0)}\,.$$

Here, the function $x_{\text{SABR}}(f, v)$ is the SABR one (1.7).

- Calculate the option price using the obtained parameters in the calculation model (its exact solution is available!)

In the example above the efficient Black volatility (implied volatility) is

$$\sigma_B = \frac{\log F_0/K}{\gamma^{-1} \log\left(\frac{\sqrt{1-2\rho\gamma\, y(F_0,v_0)+\gamma^2\, y^2(F_0,v_0)}+\gamma y(F_0,v_0)-\rho}{1-\rho}\right)}\,. \qquad (1.8)$$

One can also find the local volatility corresponding to the SABR for *all* the strikes. For this one should match

$$-\int_{F_0}^{K} \frac{\mathrm{d}F}{\sigma_L(F)} = x_{\mathrm{SABR}}(F_0, v|K) \;\; \forall K\,,$$

where we restore the dependence from the strike in the function $x_{\mathrm{SABR}}(F_0, v|K)$ replacing $y(F_0, v_0)$ by $y(F_0, v_0 \mid K)$ in (1.7). It is easy to see that

$$\begin{aligned}\sigma_L(F) &= -\left(\frac{\partial x_{\mathrm{SABR}}(F_0, v|K)}{\partial K}\bigg|_{K=F} \right)^{-1} \\ &= vF^{\beta}\sqrt{1 - 2\rho\gamma y(F_0, v_0|F) + \gamma^2 y^2(F_0, v_0|F)}\,. \end{aligned} \tag{1.9}$$

1.4 Modifications and Extensions of the SABR

1.4.1 PDE SABR Model

One of modifications of the SABR is an equivalent local volatility model (Andreasen and Huge [6], Hagan [41]), defined by an SDE

$$\mathrm{d}F_t = vF_t^{\beta}\sqrt{1 - 2\rho\gamma y(F_0, v_0|F_t) + \gamma^2 y^2(F_0, v_0|F_t)}\,\mathrm{d}W_t\,, \tag{1.10}$$

corresponding to (1.9). Clearly it is arbitrage-free (the forward has zero drift) and delivers an accurate SABR approximation for short maturities. However, it has certain disadvantages. First, its implementation requires a numerical method which is more complicated than a closed form solution. Second, the solution can strongly deviate from the SABR for long maturities. Finally, local volatility dependence on the spot F_0 makes the hedging procedure tricky—it is not evident what we should keep constant while moving the spot on the next day.

1.4.2 Negative Rates Extensions

At the time when the SABR model was introduced, positivity of the rates seemed like a reasonable and an attractive property. However, as rates became extremely low or even negative, the classical SABR could not handle this new market environment, which led to urgent need for extensions of the SABR model.

The simplest way to take into account negative rates is to shift the SABR process getting an SDE of the form

$$\mathrm{d}F_t = (F_t + s)^{\beta}\, v_t\, \mathrm{d}W_t,$$

where s is a deterministic positive shift. This moves the lower bound on F_t from 0 to $-s$. One can either include the shift into calibration parameters $(v_0, \beta, \rho, \gamma, \mathbf{s})$ or fix it prior to calibration (e.g., to 2% in case of short rates of the Swiss Franc). Both ways have drawbacks.

Calibrating the shift does not really introduce a new degree of freedom—its influence on the skew is very similar to the power β, and may result in an identification problem. On the other hand, if one selects the shift value *manually* and calibrates only the standard parameters $(v_0, \beta, \rho, \gamma)$, there is always a danger that the rates can go lower than anticipated, and one will need to change this parameter accordingly. This can result in jumps in the other SABR parameters as the calibration response to such readjustment. As a consequence, there will be jumps in the values/Greeks of trades that are dependent on the swaption or cap volatilities. To cover for potential losses in such situations, traders are likely to be asked to reserve part of their P&L. Moreover, having the swaption prices being bounded from above (due to rate being bounded from below) can lead to situations when the shifted SABR cannot attain certain market prices.

To overcome these complications, a more natural and elegant solution was developed in [12], so called Free SABR model with the following SDE,

$$\mathrm{d}F_t = |F_t|^\beta \, v_t \, \mathrm{d}W_t$$

with $0 \leq \beta < \frac{1}{2}$, and a *free* boundary. Such a model allowed for negative rates and contained a certain "stickiness" at zero. Moreover, the model satisfies norm-preserving and martingale requirements.

The authors calculated the exact option price for the *zero-correlation* case and came up with an accurate approximation for general correlations. However, the approximation accuracy can degenerate in some cases (especially those related to large strikes and high correlation).

Later on, building on the normal SABR with a free boundary,

$$\mathrm{d}F_t = v_t \, \mathrm{d}W_t,$$

a new model was proposed [13], as a *mixture* of the normal SABR and the zero-correlation Free SABR. Importantly, this model is arbitrage free and does not require any approximation in the option-pricing formula. Moreover, it contains more parameters to ensure an accurate calibration to swaptions and CMS payments. In the numerical experiments section, we demonstrate the superiority of the Mixture SABR calibration over the Shifted and Free SABR models.

1.4.3 Other Applications: SABR LMM Etc

There is the SABR process application as a term structure model; see, for example, Rebonato et al. [68] for the SABR/LIBOR Market Model or Mercurio and Morini [65] for inflation models.

Let us briefly discuss Rebonato Libor model. Given a yield curve granularity expressed in Libor dates $\{T_n\}_{n=1}^{N}$, the model follows evolution of forward Libor rates $L_n(t)$ between T_n and T_{n+1} defined as

$$dL_n(t) = \cdots + L_n(t)^{\beta_n} \, \nu_n(t) \, dW_n(t)$$

with a log-normal volatility $d\nu_n = \gamma_n \, \nu_n \, dZ_n$ such that all Brownian motions W_n and Z_n are somehow correlated. An attractive feature of this model is its simple calibration to caplets underlying the corresponding Libor rates: it is sufficient to take the caplet SABR parameters and substitute them in the forward Libor SDE's.

However, such approach will have a set of drawbacks. First, an unnatural absence of the mean-reversion of the stochastic-volatility (log-normal) results in its huge variance at 20–30Y. Second, analytical approximations for caplets can differ significantly from the simulation results for long maturities (a problem for a *term-structure* model). Moreover, the calibration to swaption smiles is problematic: (1) the model lacks free parameters (they are used for the caplets calibration); (2) analytical swaption volatility approximation is far from trivial. Calibration ability of the SABR LMM to other ATM swaptions is reduced w.r.t. the classical LMM where the volatilities are time-dependent. There are also certain numerical issues. A long term simulation of the stochastic volatility is complicated and requires a lot of time steps and paths. Moreover, pricing callable instruments (the goal of term-structure models) by regression is almost impossible: (1) a large number of states ($\sim$2 times more than in the classical LMM; (2) huge variance in the stochastic volatility for moderate-large maturities hides the information of the rates during the regression.

1.5 CMS and the SABR

Another important application of the SABR model consists of the calculation of European CMS products associated with the swap rate in hand. The CMS price is calculated via integrals of European swaption prices using the static replication formula [40]. We recall that the CMS convexity adjustment depends on the variance of the rate process. Indeed, a CMS payment equals to an expectation of the corresponding rate

$$F_t = \frac{P(t, T_0) - P(t, T_N)}{A(t)}$$

in the forward measure for the payment date T. Here $P(t, T)$ is a zero coupon bond price and $A(t)$ is the swap annuity. On the other hand, we know rate's distribution (in our case, following the SABR) in its martingale measure corresponding to the swap annuity numeraire. Thus, the CMS payment in the martingale measure is

$$\mathbb{E}\left[F_t \frac{P(t,T)}{A(t)}\right]$$

The change of measure process $P(t,T)/A(t)$ is approximated as a linear function of the rate

$$\frac{P(t,T)}{A(t)} \simeq \frac{P(0,T)}{A(0)} + \xi(F_t - F_0)$$

with a coefficient ξ often calculated assuming that the zero rate coincides the swap rate F_0. The second term in the CMS payment

$$\mathbb{E}\left[F_t \frac{P(t,T)}{A(t)}\right] \simeq F_0 \frac{P(0,T)}{A(0)} + \xi \mathbb{E}\left[(F_t - F_0)^2\right]$$

is called a convexity adjustment. It can be evaluated by the static replication formula for the process variance underlying the CMS price

$$\mathbb{E}\left[(F_T - F_0)^2\right] = 2\int_{-\infty}^{\infty} \mathrm{d}K\, \mathcal{O}(T,K). \tag{1.11}$$

where $\mathcal{O}(T,K)$ is an option time value. The standard replication [40] is valid for a positive rate F with integration starting from zero strike. In our case, the replication formula (1.11) can be used for an arbitrary support distribution which is reflected by the infinite lower bound of the integral over the strikes. The derivation is based on the following identity valid for all values of x

$$x^2 = 2\int_0^{\infty} \mathrm{d}K\, (|x| - K)^+.$$

In general, a calculation the CMS convexity adjustment procedure is done numerically integrating the option time-value for sufficiently large range of strikes. This means that the SABR swaption calculation should be robust and coherent for all the strikes which is quite problematic for the original Hagan approximation [39] valid for a moderate interval around the initial rate point F_0. In the next section, we will address the approximation accuracy in more details.

1.6 Approximation Accuracy and Its Improvements

1.6.1 Approximation Accuracy

An approximation quality of the classic Hagan formula is good for small times and moderate strike deviations from the initial rate (ATM strike). For example, for maturities larger than 10Y the error in Black implied volatility can be 1% or more

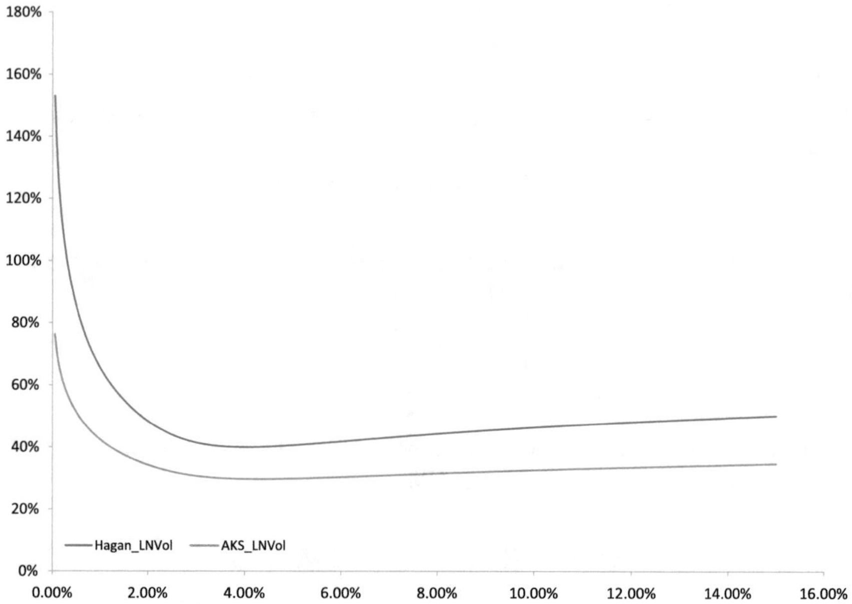

Fig. 1.2 Lognormal volatilities for SABR with parameters $F = 3\%$, $T = 10$, $\alpha = \frac{1\%}{F^{\beta}}$, $\beta = \frac{1}{2}$, $\rho = 0\%$, $\gamma = 0.53$

even for ATM values (the order of the volatility itself can be 20–30%). Moreover, one can easily observe bad approximation behavior for extreme strikes which sometimes prevents obtaining a valid probability density function: the second derivative of the option over the strikes can become negative. As we have mentioned, these undesired properties on the edges are especially dangerous for CMS calculations by static replication. We can see an example of lognormal volatility plots for the original approximation vs our new formula below in Fig. 1.2 and the probability density plots in Fig. 1.3. One can easily see that the PDF is negative and the Black volatility explodes for the original formula in the left strike tail whereas both look reasonable for our new formula.

As a poor man's solution, one can experimentally determine a valid strike range[1] for the original approximation and smoothly extrapolate the option price (with few parameters) outside of this range. Consider the main SABR usage, the implied volatility interpolation. If we have sufficient number of quotes inside the valid strike range, we can successfully calibrate the model and obtain the interpolated strikes. The extrapolation will be also determined by the valid strike. However, if the quoted strikes are also presented outside of the valid range (which is a usual case for moderate/large maturities) there can be a incoherence: the extrapolated valid range SABR can be in conflict with the quotes.

[1]This rate obviously shrinks with time going up.

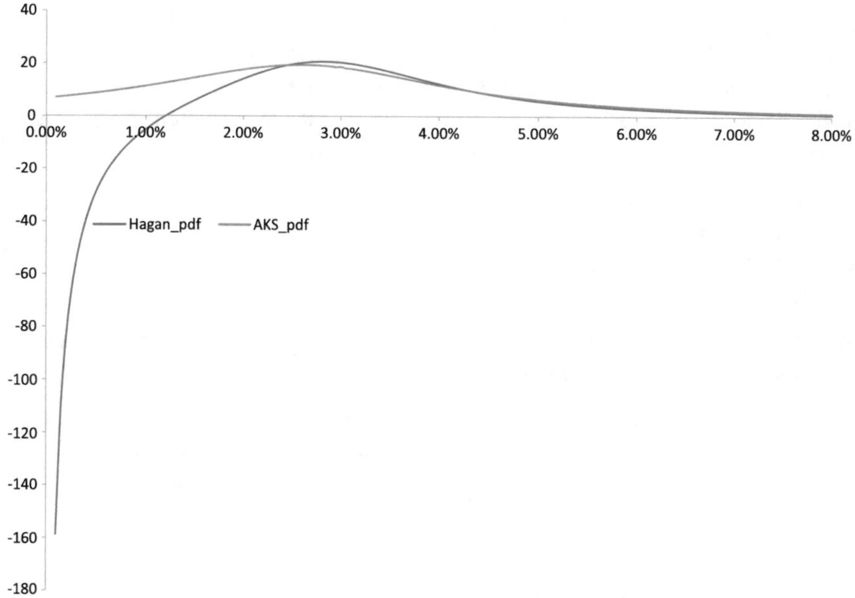

Fig. 1.3 Probability density for SABR with parameters $F = 3\%$, $T = 10$, $\alpha = \frac{1\%}{F^\beta}$, $\beta = \frac{1}{2}$, $\rho = 0\%$, $\gamma = 0.53$

In any case, a closed-from solution or at least a closed-form approximation that can be extended to the whole strike range without resorting to additional ad hoc extrapolation would present a much desired alternative to such methodologies. Below we describe such solutions paying attention to their feasibility. Namely, if a solution is presented in terms of multi-dimensional integrals, it can be too slow to be used in volatility cubes.

1.6.2 Exact Solutions

As alternatives to the Hagan approximation was concentrated on special cases where people tented to find exact solutions. Islah [50] found an option price for the zero correlation case in terms of a multi-dimensional integration. Nevertheless, a practical implementation of Islah's result for calibration was hardly possible: the final formula consisted of three-dimensional integration of special functions and was unfeasible to use in volatility cubes. Next, in [10] the authors transformed the zero correlation solution in one-dimension integral over a new special function

$$G(t, s) = 2\sqrt{2}\frac{e^{-\frac{t}{8}}}{t\sqrt{2\pi t}}\int_s^\infty du\, u\, e^{-\frac{u^2}{2t}}\sqrt{\cosh u - \cosh s} \tag{1.12}$$

This function is closely related to the McKean heat kernel on $\mathbf{H}^2$. Although being 1D integral, the function $G(t, s)$ can be extremely accurately approximated by a closed formula [10]. For example, for far strikes at 30Y an implied volatility error can be 10bps. The efficient approximation of G makes possible a usage of obtained 1D integral expression for real-time calibration procedures (the same numerical complexity as the Heston). In the list below, we specify the SABR flavors which can be exactly solved as one-dimensional integrals of the function G:

- Classic (absorbing boundary) zero-correlation [10, 11]
- Free boundary zero-correlation [12]
- Free boundary normal ($\beta = 0$) [13]
- Lognormal ($\beta = 1$) zero correlation [73]

For example, for the normal SABR $\mathrm{d}F_t = v_t \,\mathrm{d}W_t$ with some correlation ρ, the option time-value can be reduced to the following simple form

$$\begin{aligned}\mathscr{O}_N(T, K) &= \mathbb{E}\left[(F_T - K)^+\right] - (F_0 - K)^+ \qquad (1.13)\\ &= \frac{v_0}{\pi\gamma}\int_{s_0}^{\infty} \frac{G(\gamma^2 T, s)}{\sinh s}\sqrt{\sinh^2 s - (k - \rho\cosh s)^2}\mathrm{d}s,\end{aligned}$$

where

$$\cosh s_0 = \frac{-\rho k + \sqrt{k^2 + 1 - \rho^2}}{1 - \rho^2} \quad \text{for} \quad k = \frac{K - F_0}{v_0}\gamma + \rho.$$

Note that an equivalent form for the normal SABR option price first appeared in [54] in terms of a 2D integral.

1.6.3 More Accurate Approximations

In the Sect. 1.3, we described a mapping procedure which calculates option prices using an exactly solvable model that coincides with the model in hand (e.g. a general SABR model) for small times. The classic Hagan approximation uses Black-Scholes (BS) model as the mimicking one. In [10], the authors went further and considered a zero correlation SABR as a mimicking model that is obviously "closer" to the SABR than a smile-less BS. This has improved the approximation accuracy compared to the original approximation. The same procedure has been applied for a general free boundary SABR [12]. Note that the obtained approximation can be further improved using the exactly solvable normal free SABR as a control variate.

1.7 About This Book

The target audience of this book is mainly financial industry practitioners (quants, former physicists, etc.). However, it can be also interesting to mathematicians who seek intuition in the mathematical finance. In the book we pursue the following goals:

1. Unify often scattered SABR analytics in the same text
2. Provide simple and intuitive demonstrations of the main results

3. Provide mathematical formulas and their financial intuition with a support of numerical experiments
4. Make it readable for practitioners (quants, former physicists etc.) and mathematicians
5. Present numerous numerical results comparing analytics with simulation which can serve as validation benchmarks
6. Present simple and intuitive insights for such non-trivial concepts in the stochastic processes and the approximation theory as:

   ✓ Absorbing and reflecting boundaries
   ✓ Local and global marginality
   ✓ Small time asymptotic analysis
   ✓ Gyongy Lemma
   ✓ Various methods, including the mapping, that improve an approximation accuracy

   to mention a few.

In the list below we present the most interesting themes from the book chapters:

- CEV as a special case of the SABR
  - Understanding of the absorbing and reflecting boundaries using PDE's
  - Simple derivation of both absorbing and reflecting solutions
  - Conservation of the norm and marginality (local and global martingales) from the forward Kolmogorov equation
- Exactly solvable cases
  - Multiple derivations of PDF of normal free boundary SABR and a lognormal SABR
  - Simple derivation of a normal free boundary SABR option price
  - Simple derivation of a zero correlation absorbing and free boundary SABRs
- SABR option price for small times
  - Comprehensive description of the Heat-Kernel
  - Simple derivation of the SABR option price for small maturities
  - Mapping approximation and related numerical experiments
- SABR for negative rates
  - Financial intuition for negative rates
  - Comprehensive description of a free boundary
  - Simple derivation of option prices
  - Understanding of the SABR Mixing model and its relation with CMS's with numerous numerical experiments

Chapter 2
Exact Solutions to CEV Model with Stochastic Volatility

2.1 Introduction

Stochastic volatility models are used widely in interest rate modeling to match empirical evidence of stochastic volatility for interest rates and accommodate observed market skew and smile. Two of the most widely used models for European options are "Heston" [44] and "SABR" [39]. We consider the case of "constant elasticity of variance" (CEV) local volatility (as used in SABR) and general stochastic volatility [11]. Assuming zero-correlation between asset and volatility, we extend the work of [9, 10] to derive closed-form expressions for the option price in terms of the moment generating function of integrated variance or of inverse integrated variance [11]. We apply results on such functions to give an integral representation in the case that variance follows a mean-reverting square root process (as in Heston). Furthermore we show how the general framework may also be used to recover the exact solution to the SABR model obtained in [9].

Since the SABR model represents extension of the CEV model to stochastic volatility, it is natural to start our analysis with the CEV process driven by the SDE $\mathrm{d}F = VF^{\beta}\mathrm{d}W$ with the instantaneous volatility $\sigma(F) = VF^{\beta-1}$. It was introduced by Cox [20], originally for $\beta < 1$, to capture the negative correlation of volatility change with stock returns (leverage effect). For $\beta = 1$ the CEV model reduces to the geometric Brownian motion process with constant volatility, when $\beta = 0$, the process F is Brownian (Bachelier), and for $\beta = 1/2$ the model reduces to the square-root model of Cox and Ross [21]. Emanuel and MacBeth [30] extended consideration to the case $\beta > 1$, used in modeling commodity prices that exhibit increasing implied volatility skews with the volatility increasing with the strike price [32].

Despite there exists extensive literature on the subject (see [52] and literature therein), we found it appropriate to give derivation of probability density and option prices for the CEV model based on the analysis of solutions behavior close to the singular point of diffusion. We pursued two goals here: applying it later to the SABR model and pointing out some existing inaccuracies and mistakes.

A. Antonov et al., *Modern SABR Analytics*,
SpringerBriefs in Quantitative Finance,
https://doi.org/10.1007/978-3-030-10656-0_2

2.2 Transforming CEV Process into the Bessel One

We absorb constant scale parameter V into stretched time $t' = V^2 t$ and a new Brownian motion $W'_{t'} = V\mathrm{d}W_t$, then drop primes and present the driving SDE as

$$\mathrm{d}F_t = F_t^{\beta}\mathrm{d}W_t. \tag{2.1}$$

In fact, we will proceed with analysis of the squared Bessel process (BESQ) X_t, which is related to CEV F as [20]

$$X_t = \frac{F_t^{2(1-\beta)}}{(1-\beta)^2}. \tag{2.2}$$

According to Ito's lemma, process X follows the canonical dynamics of BESQ [52]

$$\mathrm{d}X_t = 2(\nu+1)\mathrm{d}t + 2\sqrt{X_t}\mathrm{d}W_t \tag{2.3}$$

with drift ν, also called index of BESQ X_t,

$$\nu = -\frac{1}{2(1-\beta)} \tag{2.4}$$

and so called dimension

$$\delta = 2 + 2\nu = \frac{1-2\beta}{1-\beta}$$

Probability density function (PDF) of BESQ process can be found by probabilistic methods [29, 33, 63] or by solving directly the forward Kolmogorov equation [31]. The probabilistic approach is based on the additivity property of Bessel processes [52], which takes place only for positive dimensions δ ($c > 0$ in Feller notations). Consideration of negative dimensions (negative c) requires more refined methods like time inversion, or changing drift and measure. Solving Kolmogorov equation with the help of the Laplace transform, as employed in the seminal paper by Feller [31], is more straightforward, even if more lengthy. Yet the way it was presented in [31] is rather complicated and can be simplified significantly.

Feller, considered the Fokker-Planck (forward Kolmogorov) equation for the random process X_t, governed by the stochastic differential equation (SDE)

$$\mathrm{d}X_t = (bX_t + c)\mathrm{d}t + \sqrt{2aX_t}\mathrm{d}W_t \tag{2.5}$$

with constant a, b, c ($a > 0$), which is used in the CIR, Heston models, to describe Bessel processes, and often called the square root process. SDE (2.5) can be reduced to the canonical BESQ SDE (2.3) by introducing the new process, time, and time changed Brownian motion,

$$\tilde{X}_{\tilde{t}} = \frac{2}{a}\mathrm{e}^{-bt}X_t,$$
$$\tilde{t} = \frac{1-\mathrm{e}^{-bt}}{b}$$
$$\mathrm{d}\tilde{W}_{\tilde{t}} = \mathrm{e}^{-bt/2}\mathrm{d}W_t,$$

where the new process $\tilde{X}_{\tilde{t}}$ (tildes dropped) follows (2.3) with dimension $\delta = \frac{2c}{a}$. Forward Kolmogorov equation for PDF of process (2.3) looks like

$$p_t = -2(1+\nu)p_x + 2(xp)_{xx} = 0. \tag{2.6}$$

Depending on ratio $\frac{c}{a}$, three different regions are distinguished in [31], the corresponding conditions for dimension $\delta = 2\frac{c}{a} = 2(\nu+1)$, index $\nu = -\frac{1}{2}(1-\beta)^{-1}$, and CEV parameter β are presented in the table,

$2 < \delta$	$0 < \nu$	$1 < \beta$
$0 < \delta < 2$	$-1 < \nu < 0$	$\beta < \frac{1}{2}$
$\delta < 2$	$\nu < -1$	$\frac{1}{2} < \beta < 1$

We provide rather simple derivation of the PDF p for all intervals of parameters. We restrict ourselves with consideration of the transition function (TF) $u(t, x \mid x_0)$, which is the conditional PDF for process X with fixed initial value x_0, in other words it is the fundamental solution of the Cauchy problem for Eq. (2.3) with initial distribution $u(0, x) = \delta(x - x_0)$.

For an arbitrary initial distribution $p_0(x)$ PDF p is being obtained as the convolution with TF,

$$p(t, x) = \int_0^\infty p(t, x \mid x')\, p_0(x')\mathrm{d}x'.$$

2.3 Solution Behavior Near Singular Point $x = 0$, Integrability, Flux

The probabilistic aspect of the problem brings additional requirements to be imposed on PDF $p(t, x)$; it must be positive, integrable, and not norm increasing. Taking these conditions into account from the very beginning along with the analysis of the solution behavior, close to the singular point $x = 0$, dramatically simplifies construction of solutions to the FK equation (2.6).

2.3.1 Flux

Flux $j(t,x)$, is an important characteristic of a solution p, its definition follows from the Kolmogorov equation (2.6) represented in the form of continuity equation

$$p_t + j_x = 0, \tag{2.7}$$

that results in the expression

$$j(t,x) = 2(\nu+1)p - 2(xp)_x = 2\nu p - 2xp_x = -2x^{1+\nu}(x^{-\nu}p)_x. \tag{2.8}$$

Flux $j(t,x)$ is measured by the amount of probability δP, crossing point x per unit time interval $\mathrm{d}t$, namely, $\delta P = j\mathrm{d}t$ (from left to right, if $j(t,x)$ is positive). Integrating equation (2.7) in some interval (a,b), we get the integral form of (2.7)

$$\frac{\partial P(X_t \in (a,b))}{\partial t} = j(t,a) - j(t,b), \tag{2.9}$$

where

$$P(X_t \in (a,b)) = \int_a^b p(t,x)\mathrm{d}x$$

is the cumulative probability (CP) to find process X_t in the interval (a,b). In particular, taking $b \to \infty$ and $a \to 0$, we get the total cumulative probability for process X_t to take any positive value, called the norm of solution

$$N(t) = P(X_t > 0) = \int_0^\infty p(t,x)\mathrm{d}x$$

(the singular point $x = 0$ is not included into N and may require a special treatment). Taking into account that flux (2.8) at infinity turns into zero due to integrability of PDF, $j(t,\infty) = 0$, we obtain from (2.9) the balance of norm in the form

$$\frac{\partial N(t)}{\partial t} = j(t,0). \tag{2.10}$$

Distribution $p(t,x)$ is called norm preserving if $N(t) = 1$ and is norm defective if $N(t) < 1$, in the latter case there arises a *finite* probability for process X_t to take the zero value,

$$P(X_t = 0) = 1 - N(t).$$

The following statements, simple, as they are, help a lot in selecting proper solutions of FK equation (2.6)

- Norm is restricted from above, $N(t) \leq 1$ and may grow only from a norm defective state;

- If the initial distribution $p(0, x)$ is normalized (and this is certainly true for transition function with $p(0, x) = \delta(x - x_0)$), then the initial norm is already at its maximum level, $N(0) = 1$, and can only decrease initially (or to stay unchanged);
- Then increase of norm (if any) may occur only after some decrease;
- Since change of norm N is related to flux at origin (2.10), the last statement can be translated to the one about flux, namely, flux at origin $j(t, 0)$ may take strictly positive values only after having taken strictly negative ones;
- As we will see below, flux $j(t, 0)$ does not change its sign. It follows then that if the flux is nonnegative at all times, $j(t, 0) \geq 0$, then it must be identically equal to zero, $j(t, 0) \equiv 0$. In such a case the total cumulative probability remains equal to one (norm preserving solution), and the singular point $x = 0$ is reflecting;
- On the opposite, if the flux is nonpositive, there are no restrictions on negative values. If $j(t, 0) < 0$, the norm $N(t)$ is decreasing, the singular point $x = 0$ is absorbing and accumulating a *finite* (and increasing) amount of probability keeping the total balance intact, $P(X_{t=0} = 0) = 1 - P(X_{t=0} > 0)$.

Finally, it is instructive to present the boundary characterization for the CEV process F_t (and for the corresponding BESQ process X_t) [23]

$\beta > 1$	$F = 0\ (X = \infty)$	natural boundary
$(\nu > 0)$	$F = \infty\ (X = 0)$	entrance
$1/2 < \beta < 1$	$F = 0\ (X = 0)$	exit
$(\nu < -1)$	$F = \infty\ (X = \infty)$	natural boundary
$\beta < 1/2$	$F = 0\ (X = 0)$	regular boundary
$(-1 < \nu < 0$	$F = \infty\ (X = \infty)$	natural boundary

2.3.2 Indexes of Solution

In the vicinity of singular point $x = 0$, let us look for solution of FK equation (2.6) in the form

$$p \sim x^{\gamma} \varphi(t, x),$$

where $\varphi(t, x)$ is a regular function of x (possessing the Taylor series). Substituting this form into (2.6), collecting terms of the leading order ($\sim x^{\gamma-1}\varphi(t, x)$), and equating coefficient at them to zero,

$$x^{\gamma-1}\varphi(t, x) \,|\quad 2(\nu + 1)\gamma - 2(\gamma + 1)\gamma = 0\,,$$

we find two roots

$$\gamma_1 = 0, \qquad \gamma_2 = \nu,$$

they are called indexes of solutions at the singular point $x = 0$. Solution with the zero index ($p \sim x^0$) is a regular function of x and is always integrable at zero, while another one ($p \sim x^\nu$) is integrable only for $\nu > -1$ (integrability is required to have a finite probability $\int p\mathrm{d}x$).

2.3.3 Flux at Origin, Solution Selection

Using definition (2.8), we calculate fluxes created at $x = 0$ by each of two types of solutions with indexes $\gamma_1 = 0$ and $\gamma_2 = \nu$. The first solution is a regular function at small x and can be expanded in Taylor series, $p_1(t, x) = C_0(t) + C_1(t)x + o(x)$. According to (2.8), it creates flux $j_1(t, x) = 2\nu C_0(t) + o(1)$ with value at origin $j_1(t, 0) = 2\nu C_0(t)$. The second solution p_2 behaves as x^ν at small x. Calculating flux j_2 it creates, by using the rightmost expression in (2.8), we see that function $x^{-\nu} p_2$ is regular, and such is its derivative $(x^{-\nu} p_2)_x$. This means that flux j_2 is of the order of $x^{1+\nu}$ and vanishes at $x = 0$ (recall that solution p_2 may be realized only if $1 + \nu > 0$ due to integrability requirement).

Next, follow separately positive and negative drifts. If drift is positive, $\nu > 0$, then $p_2 \sim x^\nu$ itself turns into zero with x, implying that p_2 contributes to neither PDF nor flux at origin, and they both are entirely due to the first part of solution, p_1

$$p(t, 0) = p_1(t, 0) = C_0(t), \qquad j(t, 0) = j_1(t, 0) = 2\nu C_0(t).$$

Function $C_0(t)$, as describing probability density, must be always positive (at least nonnegative), then flux, $j(t, 0) = 2\nu C_0(t)$ is positive as well (at $\nu > 0$). This situation has just been discussed with conclusion made that nonnegative flux means zero flux, $j(t, 0) \equiv 0$, which in turn implies that $C_0(t) \equiv 0$. With C_0 vanished, regular solution $p_1(t, x)$ should start with the first order term $\sim x$. But $\gamma = 1$ is not a characteristic index of Eq. (2.6), and in absence of C_0, coefficient at the leading term ($\sim C_1$) can not be balanced otherwise than putting $C_1 \equiv 0$. Consecutively we obtain $C_2 \equiv 0$, $C_3 \equiv 0$, etc., the whole solution p_1 is merely equal to zero. To be exact, substitution of the Taylor series $\sum C_k x^k/k!$ into Eq. (2.6) generates the recursion relation $2C_{k+1} = \dot{C}_k/(k + 1 - \nu)$, which results in identical zeros for all coefficients C_k, if $C_0 \equiv 0$ (special cases of ν equal to positive integers need some additional consideration, but do not change the result). Solution $p_1(t, x)$ which behaves regularly ($\sim x^0$) at small x, can not be realized at positive drifts, $\nu > 0$, as it would result in a wrong (outward) flux at origin, thus violating conservation of probability.

For strong negative drifts, $\nu < -1$, there are no such contradictions, and at origin, $x \to 0$, positive PDF $C_0(t) > 0$, may coexist 'peacefully' with negative (inward) flux $j(t, 0) = 2\nu C_0 < 0$.

The following conclusions can be made

- For positive drifts, $\nu > 0$ ($c > a > 0$ by Feller [31]), only solution $p_2 \sim x^\nu$ may be realized. It creates zero flux and zero PDF at origin and is norm preserving. Singular point $x = 0$ is repulsing (unattainable for process X_t);

- For strongly negative drifts, $\nu \leq -1$ $(c < 0)$ only solution $p_1 \sim x^0$ may be realized (integrability). It creates negative (inward) flux at origin, thus causing accumulation of probability at $x = 0$, being the absorbing singular point;
- In both cases, the transition function is unique, and unique is the solution of the Cauchy problem with an arbitrary (yet positive and normalized) initial distribution;
- At moderate negative drifts, $-1 < \nu \leq 0$ $(0 < c < a)$, there is no general criteria to choose between two solutions, and they both can be realized. Transition function (or solution of the initial value problem) can not be found uniquely without additional information about the dynamics near the singular point $x = 0$.

We provide some heuristic arguments in favor of solutions selection. At very large positive drifts, the drift term in SDE (2.3) will immediately throw process X away, if X accidentally approaches zero. This means that the region near zero is 'populated' scarcely, therefore only the reflecting solution $p_2 \sim x^{\nu}$ can be realized at large drifts. When drift decreases, we stick with this solution due to continuity of the probability law with respect to drift. When drift reaches bifurcation point $\nu = 0$, where two solution indexes coincide, another, absorbing solution $p_2 \sim x^0$ may emerge continuously. As drift becomes negative, both solutions coexist until drift reaches value $\nu = -1$ where two solutions merge. This time it is due to the special feature of modified Bessel functions $I_{-1}(z) \equiv I_1(z)$ that are involved in the probability density of Bessel processes (below). The reflecting solution 'dies' at this point, and only the absorbing one exists at $\nu < -1$.

2.4 Laplace Transform

Consider the Laplace transform (LT) of PDF p with respect to variable x,

$$F(t, q) = \int_0^{\infty} p(t, x) \mathrm{e}^{-qx} \mathrm{d}x$$

(in the case process X_t is not accumulated at zero, LT F represents the moment generating function of the process, $F = \mathbb{E}[\mathrm{e}^{-qX_t}]$). Since coefficients of the FK equation (2.6) are linear in x, the Laplace transform converts (2.6) into the first order differential equation for F with respect to parameter q. On the downside, the Laplace transform of derivatives contains boundary terms, thus leading in general to a non-homogeneous differential equation. Indeed, taking the LT of FK in the form (2.7), we get

$$F_t + q \int_0^{\infty} j(t, x) \mathrm{e}^{-qx} \mathrm{d}x = j(t, 0),$$

then using expression (2.8) to calculate LT of flux $j(t, x)$, we obtain the equation for the Laplace transform $F(t, q)$

$$F_t + 2(\nu + 1)qF + 2q^2F_q = j(t, 0), \tag{2.11}$$

with initial condition

$$F(0, q) = \int_0^\infty p(0, x)\mathrm{e}^{-qx}\mathrm{d}x = \int_0^\infty \delta(x - x_0)\mathrm{e}^{-qx}\mathrm{d}x = \mathrm{e}^{-qx_0}. \tag{2.12}$$

The solution $F(t, q)$ still depends functionally on an unknown function $j(t, 0)$ (denoted $f(t)$ in the Feller paper) that poses a real difficulty. In general, one should write PDF $p(t, x)$ as inverse LT of F, and calculate flux $j(t, 0)$ at $x = 0$ according to (2.8), thus arriving to an integral equation for $j(t, 0)$. When (and if) $j(t, 0)$ is found, one would return to F and restore PDF p explicitly. Feller [31] managed to do this for negative drifts, $\nu < 0$, with lengthy formal manipulations required. The case of positive drifts $\nu > 0$ required series of nontrivial lemmas to prove that $j(t, 0)$ must be equal to zero.

Instead, our simple analysis of the behavior near singular point $x = 0$ allows to avoid these difficulties. As shown in the previous section, the probabilistically sound solution for positive drifts, $\nu > 0$, must not possess flux at origin, $j(t, 0) \equiv 0$, so that Eq. (2.11) for $F(t, q)$ becomes homogeneous.

At negative drifts, $\nu < -1$, dealing with non zero flux at zero can be avoided as well by exploiting the method that is often used for ordinary differential equations. Imagine we have the singular point at $x = 0$ with two indexes of solutions $\gamma_+ > \gamma_-$, such that substitution $p \sim x^\gamma \tilde{p}$ with either of γ converts our equation into an one with linear coefficients. Trying to construct the solution with *higher* index by applying LT, we make substitution with the *lower* index, $p \sim x^{\gamma_-}\tilde{p}$. Underline, this does not suggest switching to another solution, such substitution means only that new function $\tilde{p}$ behaves like $x^{\gamma_+ - \gamma_-}$, thus turning into zero at $x = 0$. In doing so, we get the best of two worlds - on the one hand, the lower index γ_- has the same effect, resulting in linear coefficients, on the other hand, taking the Laplace transform of equation for function $\tilde{p}$ will not generate boundary terms since $\tilde{p}_{|x=0} = 0$.

We notice in our case that for positive drifts the higher index is $\gamma = \nu$, the lower one is $\gamma = 0$, and no substitution is needed so that LT may be applied directly to Eq. (2.6), exactly as we have done.

For negative drifts, $\gamma = 0$ becomes the higher index, and $\gamma = \nu < 0$ is the lower one. In this case we make substitution

$$p(t, x) = \left(\frac{x}{x_0}\right)^\nu \tilde{p}(t, x) \tag{2.13}$$

suggesting that $\tilde{p}$ behaves like $x^{-\nu}$ at small x, and obtain from (2.6) the following equation for $\tilde{p}$,

$$\tilde{p}_t + 2(1 - \nu)\tilde{p}_x - 2(x\tilde{p})_{xx} = 0 \tag{2.14}$$

Remarkably, the last equation (ν is negative here) coincides with the original one (2.6) for positive drift $\nu' = -\nu = |\nu|$. Moreover, both the expected behavior $\tilde{p} \sim$

$x^{-\nu} = x^{\nu'}$ at small x and the initial distribution $\tilde{p}_{|t=0} = \delta(x - x_0)$ are the same as those of the sought fundamental solution of (2.6) with positive drift $\nu' = -\nu$. We conclude that $\tilde{p}$ of (2.14) merely coincides with corresponding p of (2.6).

Summarizing. If $p^{(\nu)}(t, x \mid x_0)$ is the transition function for *positive* drifts ν, then its Laplace transform $F(t, q)$ solves homogeneous equation (2.11) (with zero flux $j(t, 0)$). Solution $p^{(\nu)}$ may be extended to negative drifts $-1 < \nu < 0$. For *negative* drifts, $\nu \leqslant -1$, the transition function is expressed as

$$p_a^{(-|\nu|)}(t, x \mid x_0) = \left(\frac{x}{x_0}\right)^{-|\nu|} p_r^{(|\nu|)}(t, x \mid x_0) \tag{2.15}$$

(subscripts a and r mean absorbing and reflecting). This solution (which is regular at $x = 0$) may also be extended into interval $-1 < \nu < 0$.

The last relation between reflecting and absorbing solutions can also be obtained probabilistically by changing drift and measure of process X_t (2.3) [33, 63].

2.5 Probability Distributions

2.5.1 *Solution for Laplace Transform F*

For positive drifts, Laplace transform $F(t, q)$ of PDF p solves homogeneous equation (2.11) (with $j(t, 0) = 0$), which can be simplified further by combining two last terms and multiplying the whole equation by $q^{\nu+1}$,

$$\frac{\partial}{\partial t}(q^{\nu+1} F) + 2q^2 \frac{\partial}{\partial q}(q^{\nu+1} F) = 0. \tag{2.16}$$

The last one is readily integrated by the method of characteristics. Characteristic equation of (2.16) looks like $\mathrm{d}t = \frac{\mathrm{d}q}{2q^2}$ and integrates to

$$t + \frac{1}{2q} = \frac{1}{2q_0} \tag{2.17}$$

Constant of integration is chosen in such a way that $q = q_0$ at $t = 0$ (in mechanics q and q_0 would be called Eulerian and Lagrangian coordinates). After passing from variables (t, q) to (t, q_0), (2.16) becomes trivial $\left(\frac{\partial}{\partial t}\right)_{q_0} (q^{\nu+1} F) = 0$, its general solution is

$$q^{\nu+1} F(t, q) = g(q_0)$$

where function $g(q_0)$ is determined by the initial condition for F (2.12) (also taking into account that $q = q_0$ at $t = 0$)

$$g(q_0) = q_0^{\nu+1} F(0, q_0) = q_0^{\nu+1} \mathrm{e}^{-q_0 x_0}$$

As the last step, we express q_0, as defined in (2.17), in terms of t and q and obtain finally

$$F(t, q) = \frac{1}{(1+2tq)^{\nu+1}} \mathrm{e}^{-qx_0/(1+2tq)} \tag{2.18}$$

The probabilistic derivation of $F(t, q)$, which is also the moment generating function (MGM), is presented in [29, 69], based on the multiplicative property of MGM, which, in turn, follows from the additivity property of squared Bessel processes.

2.5.2 *Calculating Probability Distributions*

PDF p is now restored as the inverse Laplace transform

$$p(t, x \mid x_0) = \frac{1}{2\pi \mathrm{i}} \int_{C_B} \frac{\mathrm{e}^{qx - qx_0/(1+2tq)}}{(1+2tq)^{\nu+1}} \mathrm{d}q, \tag{2.19}$$

where C_B is Bromwich contour $\Re q = const > 0$. By changing variable of integration to $q' = (\frac{x}{x_0})^{1/2}(1+2tq)$, this integral becomes an integral representation for the modified Bessel function of the order ν [74].

We prefer to derive the expression (2.25) for PDF p from its LT $F(t, q)$ without involving complex integrals; following [29], we use instead properties of the well known gamma distribution $g(\lambda, z)$ and its LT F_g,

$$g(x, \nu+1, 2t) = \frac{1}{2t\,\Gamma(\nu+1)} \mathrm{e}^{-\frac{x}{2t}} \left(\frac{x}{2t}\right)^{\nu}, \tag{2.20}$$

$$F_g(\nu+1, q) = \int_0^\infty g(\cdot)\mathrm{e}^{-qx}\mathrm{d}x = \frac{1}{(1+2tq)^{\nu+1}}. \tag{2.21}$$

For brevity, we use below notation $M(q) = \frac{1}{(1+2tq)}$. Notice that if $x_0 = 0$, the exponential in (2.18) disappears, and LT $F(t, q)$ coincides with $F_g = M(q)^{\nu+1}$, accordingly, pdf p coincides with g (2.20). This gives a hint to rearrange the exponential and expand in series of $M(q)$ as

$$F(t, q) = \frac{1}{(1+2tq)^{\nu+1}} \mathrm{e}^{-\frac{x_0}{2t} + \frac{x_0}{2t} M(q)} = \mathrm{e}^{-\frac{x_0}{2t}} \sum_{n=0}^{\infty} \frac{x_0^n}{(2t)^n n!} M(q)^{n+\nu+1}.$$

As seen from (2.21, 2.20), factor $M(q)^{n+\nu+1}$ in each term represents Laplace transform of the gamma distribution g with parameter $\nu' = \nu + n$, standing for ν, thus rendering LT $F(t, q)$, as a sum of Laplace transforms. Inverting the series term by term, we present PDF p as the sum of weighted gamma distributions

$$p(t, x \mid x_0) = \mathrm{e}^{-\frac{x_0}{2t}} \sum_{n=0}^{\infty} \frac{x_0^n}{(2t)^n n!} g(x, n+\nu+1, 2t) \tag{2.22}$$

$$= \frac{1}{2t} \mathrm{e}^{-\frac{(x+x_0)}{2t}} \left(\frac{x}{2t}\right)^{\nu} \sum_{n=0}^{\infty} \frac{(x x_0)^n}{n! \Gamma(n+\nu+1)(2t)^{2n}}. \tag{2.23}$$

The last sum is expressed through the modified Bessel function $\sum(\cdot) = (z/2)^{-\nu} I_\nu(z)$ with $z = \sqrt{x x_0}/2t$, as follows from the definition of I_ν [1] in the form of the series

$$I_\nu(z) = (z/2)^\nu \sum_{n=0}^{\infty} \frac{(z/2)^{2n}}{n!\, \Gamma(n+\nu+1)}. \tag{2.24}$$

Finally, we obtain the following PDF for positive drifts [63]

$$p_r^{(\nu)}(t, x \mid x_0) = \frac{1}{2t} \mathrm{e}^{-\frac{x+x_0}{2t}} \left(\frac{x}{x_0}\right)^{\nu/2} I_\nu\left(\frac{\sqrt{x x_0}}{t}\right). \tag{2.25}$$

This is the unique transition function for positive drifts $\nu \geq 0$ ($\delta \geq 2$); in this case the zero x point is unattainable. Solution (2.25) can also be realized as the reflecting transition function in the interval ($-1 < \nu < 0$) ($0 < \delta < 2$), when point $x = 0$ can be reached and is instantaneously reflecting.

As shown above (2.15) for negative drifts, $\nu < 0$ ($\delta < 2$), the absorbing transition function is obtained from the reflecting one in the form [63]

$$p_a^{(-|\nu|)}(t, x \mid x_0) = \frac{1}{2t} \mathrm{e}^{-\frac{x+x_0}{2t}} \left(\frac{x}{x_0}\right)^{-|\nu|/2} I_{|\nu|}\left(\frac{\sqrt{x x_0}}{t}\right). \tag{2.26}$$

Notice that in the interval ($-1 < \nu < 0$) ($0 < \delta < 2$), the condition, that the process is stopped upon the first hitting zero, is imposed externally, while at $\nu < -1$ ($\delta < 0$) the singular point $x = 0$ is inherently absorbing, and the process can not exist beyond the hitting time.

Functionally expressions (2.25) and (2.26) differ only by the order of the Bessel functions; for reflecting solutions the order coincides with drift ν (whether positive or negative), and for absorbing solutions the order is equal to drift with minus $(-\nu) = |\nu|$.

2.5.3 *Properties of Reflecting and Absorbing Distributions*

Symmetries. One has been already used to construct the absorbing solution (2.15), recall

$$p_a^{(-|\nu|)}(t, x \mid x_0) = \left(\frac{x}{x_0}\right)^{-|\nu|} p_r^{(|\nu|)}(t, x \mid x_0). \tag{2.27}$$

Another one follows from rewriting factor $(x/x_0)^{-|\nu|/2}$ in (2.26) as $(x_0/x)^{|\nu|/2}$ and comparing with (2.25)

$$p_a^{(-|\nu|)}(t, x \,|\, x_0) = p_r^{(|\nu|)}(t, x_0 \,|\, x). \tag{2.28}$$

Series Expansions. We refer to [70] for series expansion of probability densities in terms of gamma distributions (in fact such expansion for reflecting solution is already given by (2.23)).

Boundary Values. As expected, the reflecting pdf behaves at small x like x^ν and creates no flux at origin. The absorbing pdf is a regular function of x and creates at origin flux (2.8)

$$j(t, 0) = -2|\nu| p(t, 0 \,|\, x_0) = -\frac{1}{t\,\Gamma(|\nu|)} e^{-\frac{x_0}{2t}} \left(\frac{x_0}{2t}\right)^{|\nu|}. \tag{2.29}$$

Small Time and Initial Distribution. Both solutions reveal the same behavior at small times $t \ll \sqrt{x x_0}$. Using asymptotic of Bessel functions [1], $I_\nu(x) \approx \frac{e^x}{\sqrt{2\pi x}}$, we get

$$p(t, x \,|\, x_0) = \tfrac{1}{2} \left(\frac{x}{x_0}\right)^{\nu/2} (x x_0)^{-1/4} \frac{1}{\sqrt{2\pi t}} e^{-\frac{(\sqrt{x} - \sqrt{x_0})^2}{2t}}.$$

With time tending to zero it turns into

$$p(0, x \,|\, x_0) = \tfrac{1}{2\sqrt{x}} \delta(\sqrt{x} - \sqrt{x_0}) = \delta(x - x_0).$$

Norm Preservation and Norm Defect. Reflecting solution (2.25) is normalized to one as immediately follows from integrating its series expansion (2.23) term by term and then summing up the resulting integrals. Another situation takes place with the absorbing pdf (2.26). It is convenient now to integrate equation (2.10) for the norm with flux given by (2.29)

$$\frac{\partial N(t)}{\partial t} = j(t, 0) = -\frac{1}{t\,\Gamma(|\nu|)} e^{-\frac{x_0}{2t}} \left(\frac{x_0}{2t}\right)^{|\nu|},$$

$$N(t) = P(X_t > 0) = 1 - G\left(|\nu|, \frac{x_0}{2t}\right), \qquad G(\lambda, z) = \frac{\Gamma(\lambda, z)}{\Gamma(\lambda)}, \tag{2.30}$$

where

$$\Gamma(\lambda, z) = \int_z^\infty e^{-z} z^{\lambda - 1} dz, \qquad \gamma(\lambda, z) = \int_0^z e^{-z} z^{\lambda - 1} dz \tag{2.31}$$

are the upper and lower incomplete gamma functions [1]. Balance of probability is accumulated at zero (mass probability) [20, 23]

$$P(X_t = 0) = 1 - N(t) = G\left(|\nu|, \frac{x_0}{2t}\right). \tag{2.32}$$

2.5.4 Cumulative Probabilities

We again refer to [70] for series expansion of cumulative probabilities in terms of cumulative gamma probabilities and gamma distributions.

BESQ and Chi-Square Distributions. PDF of the χ^2 distribution [1] with variable x, dimension $\delta = 2\nu + 2 > 0$, and noncentrality parameter y is defined as

$$p_{\chi^2}(x, \delta, y) = \tfrac{1}{2}\mathrm{e}^{-\frac{x+y}{2}} \left(\frac{x}{y}\right)^{\nu/2} I_\nu\left(\sqrt{xy}\right). \tag{2.33}$$

The cumulative probability and the complementary probability are

$$\chi^2(x, \delta, y) = \int_0^x p_{\chi^2}(x', \delta, y)\mathrm{d}x', \tag{2.34}$$

$$Q(x, \delta, y) = \int_x^\infty p_{\chi^2}(x', \delta, y)\mathrm{d}x'. \tag{2.35}$$

If noncentrality $y = 0$, chi square distribution reduces to gamma distribution

$$p_{\chi^2}(x, 2+2\nu, 0) = \frac{\mathrm{e}^{-x/2}}{2\,\Gamma(\nu+1)}\left(\frac{x}{2}\right)^\nu, \tag{2.36}$$

accordingly, cumulative probabilities are expressed through the lower and upper incomplete gamma functions (2.31)

$$\chi^2(x, 2+2\nu, 0) = \frac{\gamma(\nu+1, x/2)}{\Gamma(\nu+1)}, \qquad Q(x, 2+2\nu, 0) = \frac{\Gamma(\nu+1, x/2)}{\Gamma(\nu+1)}. \tag{2.37}$$

Underline that due to the integrability condition at $x = 0$, probability $\chi^2(\cdot)$ is defined only at positive dimensions $\delta > 0$ ($\nu > -1$), in which case $\chi^2 + Q = 1$. On the other hand, Q (2.35) and p_{χ^2} (2.33) exist as formal functions of their arguments x, δ, y at any dimensions δ, though at $\delta < 0$ ($\nu < -1$) they do not anymore bear probabilistic meaning.

Obviously, the reflecting PDF (2.25) represents the time scaled χ^2 distribution,

$$p_r^{(\nu)}(t, x \,|\, x_0) = \frac{1}{t} p_{\chi^2}(x/t, 2+2\nu, x_0/t)$$

with cumulative probability

$$P_r^{(\nu)}(X_t < x \mid x_0) = \chi^2(x/t,\ 2+2\nu,\ x_0/t). \tag{2.38}$$

As far as the absorbing PDF (2.26) is concerned, it is certainly not a chi-square PDF. But due to symmetry (2.28), it may be treated as a chi-square density with switched x and x_0

$$p_a^{(-|\nu|)}(t, x \mid x_0) = p_r^{(|\nu|)}(t, x_0 \mid x) = \frac{1}{t} p_{\chi^2}(x_0/t, 2 + 2|\nu|, x/t). \tag{2.39}$$

In this case, computing cumulative probabilities (and option values) requires integration of p_{χ^2} over the non-centrality parameter. There exists relation, due to Schroder [70], between integral of p_{χ^2} over non-centrality parameter and function Q (2.35) at smaller dimension

$$\int_y^\infty p_{\chi^2}(x, \delta, y')\mathrm{d}y' + \int_x^\infty p_{\chi^2}(x', \delta - 2, y)\mathrm{d}x' = 1, \tag{2.40}$$

which holds for any dimension δ. If and only if $\delta > 2$, it may be rewritten in the form

$$\int_y^\infty p_{\chi^2}(x, \delta, y')\mathrm{d}y' = \chi^2(x, \delta - 2, y).$$

Equality (2.40) was derived in [70] by rearranging infinite series of gamma distributions. In Appendix we provide a slightly different derivation based on [55]. Equality (2.40) was derived in [70] by rearranging infinite series of gamma distributions, also in [55] in a somewhat simpler way.

Returning to the absorbing PDF, its cumulative probability is obtained by integrating (2.39) and using (2.40)

$$P_a^{(-|\nu|)}(X_t > x \mid x_0) = \chi^2(x_0/t,\ 2|\nu|,\ x/t), \tag{2.41}$$

it reveals norm defect (2.30), $N_a(t) = P_a(X_t > 0) < 1$.

2.6 Back to CEV Model

Recall that CEV process F obeys the SDE (2.1)

$$\mathrm{d}F_t = F_t^\beta \mathrm{d}W_t.$$

The new process

$$X_t = \frac{F_t^{2(1-\beta)}}{(1-\beta)^2} = 4\nu^2 F_t^{-1/\nu} \tag{2.42}$$

follows the dynamics of the squared Bessel process

$$\mathrm{d}X_t = 2(\nu + 1)\mathrm{d}t + 2\sqrt{X_t}\mathrm{d}W_t' \qquad W_t' = \mathrm{Sign}(1-\beta)W_t$$

with drift $\nu = -\frac{1}{2(1-\beta)}$ and dimension $\delta = 2 + 2\nu$. The original CEV process F_t is expressed through the BESQ process X_t as

$$\frac{F_t}{F_0} = \left(\frac{X_t}{x_0}\right)^{-\nu}.$$

We introduce following notations for Bessel variables, associated with their CEV counterparts

$$\{x, x_0, x_K\} = 4\nu^2\{F, F_0, K\}^{-1/\nu} \tag{2.43}$$

with x_K to be used below in pricing options, struck at K.

2.6.1 Martingale Properties of the CEV Process

Expectation of F_t

$$\mathbb{E}[F_t] = F_0\mathbb{E}\left[\left(\frac{X_t}{x_0}\right)^{-\nu}\right]$$

is found by integrating BESQ PDFs (2.25) or (2.26) with weight $(x/x_0)^{-\nu}$. Express these products through p_{χ^2} (2.34)

$$(x/x_0)^{|\nu|}\, p_a^{(-|\nu|)}(t, x, x_0) = \frac{1}{t} p_{\chi^2}(x/t, 2 + 2|\nu|, x_0/t), \tag{2.44}$$

$$(x/x_0)^{-\nu}\, p_r^{(\nu)}(t, x, x_0) = \frac{1}{t} p_{\chi^2}(x_0/t, 2 + 2\nu, x/t). \tag{2.45}$$

For the absorbing PDF, integration of (2.44) gives one, meaning $\mathbb{E}[F_t] = F_0$ (notice that the probability mass at zero does not contribute because $x^{|\nu|} = 0$ at $x = 0$).

For the reflecting PDF, we integrate over non centrality parameter, using symmetry (2.40) with $y = 0$ and then the central chi-square probability (2.37)

$$\frac{\mathbb{E}[F_t]}{F_0} = 1 - Q(x_0/t, 2\nu, 0) = 1 - \frac{\Gamma(\nu, x_0/t)}{\Gamma(\nu)}.$$

Despite the last expression is valid in the whole interval where the reflecting solution exists $(-1 < \nu)$, properties of process F_t are quite different for $(0 < \nu)$ and for $(-1 < \nu < 0)$. Notice that gamma function $\Gamma(\nu)$ is positive at $0 < \nu$, but negative at $-1 < \nu < 0$, as $\Gamma(-|\nu|) = \frac{\Gamma(1-|\nu|)}{(-|\nu|)} < 0$ (incomplete $\Gamma(\nu, x_0/t)$ is always positive).

Summarizing, we observe that CEV process F_t is

- a martingale at $\beta < 1$ $(\nu < 0)$ and absorbing boundary
- a strict local martingale and a sub-martingale at $\beta < 1/2$ $(-1 < \nu < 0)$ and reflecting boundary
- a strict local martingale and a super-martingale at $1 < \beta$ $(0 < \nu)$ – reflecting boundary

2.6.2 Option Pricing Through Chi Square Distributions

We collect below option values for the CEV model, making first some preliminary comments; we use for computation associated Bessel variables x (2.43) and transform all integrands into p_{χ^2} densities (2.33); we use function Q (2.35) only when χ^2 does not exist ($\delta < 0$); otherwise ($\delta > 0$) we use only probability χ^2 (2.34) and express complementary Q as $1 - \chi^2$.

The call option value with strike K is evaluated as the expectation of payoff

$$C = F_0 \mathbb{E}\left\{\frac{F}{F_0}\mathbf{1}_{F>K}\right\} - K\mathbb{E}\left\{\mathbf{1}_{F>K}\right\} \tag{2.46}$$

with relation between call and put values looking in general like

$$C - P = \mathbb{E}[F] - K.$$

It turns into the canonical call-put parity $C - P = F_0 - K$ only for absorbing PDF, when process F_t is a martingale.

The integrand in the first term in (2.46) is given by (2.44) for absorbing PDF or by (2.45) for reflecting one (with further use of symmetry (2.40)). The second term in (2.46) integrates into the proper cumulative probability.

1. For the **absorbing boundary** at $F = 0$ ($x = 0$) and $\beta < 1$ ($\nu < 0$) the call value is

$$C = F_0\left[1 - \chi^2(x_K/t,\ 2 + 2|\nu|,\ x_0/t)\right] - K\chi^2(x_0/t,\ 2|\nu|,\ x_K/t), \tag{2.47}$$

which is the same, or equivalent to given in many sources [20, 48, 52, 61, 70]. Put and call values obey the put-call parity, on account of process F_t being a martingale.

Pricing for reflecting boundaries requires separate consideration for $\nu > 0$ ($\beta > 1$) and for $-1 < \nu < 0$ ($\beta < 1/2$). Despite PDF for BESQ process X_t is the same, properties of CEV process F_t, cumulative probabilities, and option values are completely different. In particular, reflecting point $x = 0$ corresponds to $F = 0$ for $\beta < 1$, but to $F = \infty$ for $\beta > 1$, the in the money condition for a call option, $F > K$, translates into $x > x_K$ for $\beta < 1$, but into $x < x_K$ for $\beta > 1$. This reversion of conditions was ignored in [50] and, regretfully, in monograph [52], where substantial corrections are needed to the section on CEV model with $\beta > 1$.

2. The **reflecting boundary** at $F = 0$ ($x = 0$) and $\beta < 1/2$ ($-1 < \nu < 0$) is rarely covered, as the absorbing bounadary is more coherent to the CEV model at $\beta < 1$. However, we need the technical result to compose the free CEV model in Chap. 5. The call value is

$$C = F_0[1 - Q(x_0/t,\ -2|\nu|,\ x_K/t)] - K\left[1 - \chi^2(x_K/t,\ 2 - 2|\nu|,\ x_0/t)\right]. \tag{2.48}$$

It will be wrong to write here χ^2 instead of $1 - Q$ [50, 55] because χ^2 merely does not exist at $\delta = -2|\nu| < 0$.

3. The **reflecting boundary** at $F = \infty$ ($x = 0$) at $\beta > 1$ ($\nu > 0$).

We do not continue with this case later in the book, but place some discussion here. Recall that process F_t is now a strict local martingale and a super-martingale,

$$\mathbb{E}[F_t] = F_0\left(1 - G(\nu, \tfrac{x_0}{2t})\right) \qquad \left(G(\nu, z) = 1 - \tfrac{\Gamma(\nu,z)}{\Gamma(\nu)}\right) \tag{2.49}$$

and that integration over x in (2.46) is run now from 0 to x_K. The call option value is given by the expected value of payoff [45, 59]

$$C = F_0\left(1 - G(\nu, \frac{x_0}{2t}) - \chi^2(x_0/t,\ 2\nu, \% x_K/t)\right) - K\chi^2(x_K/t,\ 2+2\nu,\ x_0/t). \tag{2.50}$$

A different valuation was presented in the original [30] and later works [2, 48, 50, 70]

$$\tilde{C} = F_0\left(1 - \chi^2(x_0/t,\ 2\nu, \% x_K/t)\right) - K\chi^2(x_K/t,\ 2+2\nu,\ x_0/t). \tag{2.51}$$

Though it was implied incorrectly that price $\tilde{C}$ is the risk-neutral expectation of the payoff (which it is not, C is), both $\tilde{C}$ and C solve the derivative valuation equation with the CEV volatility $\sigma(F) = F^{\beta-1}$

$$\partial_t V = \tfrac{1}{2}\sigma^2(F_0)F_0^2\partial_{F_0}^2 V.$$

Presence of two different prices marks CEV as the model with 'bubbles' [45, 51]. The risk-neutral expectations $\mathbb{E}[F_t]$ and C are termed fundamental (or fair) asset and option prices, while F_0 and $\tilde{C}$ – as market prices. The difference $F_0 G$ between each two, the same for the asset and call option, is the bubble. For the discussion on various aspects of delicate pricing with bubbles see [45, 51] and literature therein.

We lean toward a practical view, declared also in [45], that bubbles in models should be avoided. It is impossible, however, to avoid the CEV model with $\beta > 1$ since the asset behavior consistent with the CEV SDE (2.1) was observed in practice (the inverse leverage effect in commodities markets).

The remedy was proposed by Andersen and Andreasen [2] who termed the limited CEV process (LCEV) (for an elaborate description see also [23]). The LCEV process F_t switches to a geometric Brownian motion when F_t exceeds some large threshold F^*, accordingly, the LCEV local volatility is $\sigma = (\min\{F, F^*\})^{\beta-1}$. LCEV process F_t becomes a martingale, bubbles $F_0 G$ disappear, the standard put-call parity is restored, and the market call value $\tilde{C}$ (2.51) prevails after the limit $F^* \to \infty$ is taken. Still, pricing (2.51) does not come without restrictions. It was observed in [30, 45] that the call values $\tilde{C}$ do not tend to zero as the strike K tends to infinity. This counter-factual property was considered in [30, 45] as a trouble for the valuation (2.51) as a whole, in fact it only imposes some limitations on its applicability. Recall, that (2.51) is the risk-neutral valuation only in LCEV model, but not in CEV. Like in any model with cutoff, limit $F^* \to \infty$ must be taken after all other computations are completed, implying that sending strike K to infinity after F^* is simply not allowed in LCEV. Taking limits in the proper order, $F^* \to \infty$ after $K \to \infty$, leads to the correct, Black-Scholes zero call value. In practice, this means that the call price $\tilde{C}$ (2.51), as a function of strike, is valid only up to some large, but finite strike value K^*, which should be determined empirically.

Returning to our working cases 1. and 2., an important expansion of option values at small strikes and/or spots is considered in Chap. 5 in connection with the free CEV model.

2.7 Alternative Expressions for CEV Option Values

We consider here $\beta < 1$ ($\nu < 0$) for the absorbing PDF and $\beta < 1/2$ ($-1 < \nu < 0$) for the reflected PDF. We start with the general expression for the call option value

$$C(t, K, F_0) = \int (F - K)^+ p_F(t, F \mid F_0)\, \mathrm{d}F,$$

differentiate it with respect to time, use the FK equation

$$\partial_t p_F = \frac{1}{2}\partial_F^2 (F^{2\beta} p_F), \tag{2.52}$$

and integrate by parts to get

$$\partial_t C(t, K, F_0) = \frac{K^{2\beta}}{2} p_F(t, K \mid F_0), \tag{2.53}$$

where the PDF $p_F(t, F \mid F_0)$ of the underlying F_t is related to the PDF $p_x(t, x \mid x_0)$ of the associated BESQ process X_t as

$$p_F \mathrm{d}F = p_x \mathrm{d}x = p_x \frac{\mathrm{d}x}{\mathrm{d}F}\mathrm{d}F = 2\frac{F^{1-2\beta}}{1-\beta} p_x \mathrm{d}F.$$

Next, using relations between ν and β (2.4), F and x (2.2), and probability densities (2.26) (absorbing), or (2.25) (reflected), we get

$$\partial_t C(t, K, F_0) = |\nu| \frac{\sqrt{K F_0}}{t} \mathrm{e}^{-\frac{x_K + x_0}{2t}} I_{\pm|\nu|}\left(\frac{\sqrt{x_K x_0}}{t}\right). \tag{2.54}$$

Integration generates the option time value

$$\mathcal{O}(t, K, F_0) = C(t, K, F_0) - (F_0 - K)^+$$

in the form

$$\mathscr{O}(t, K, F_0) = |\nu|\sqrt{KF_0}\int_0^t e^{-\frac{x_K+x_0}{2t'}} I_{\pm|\nu|}\left(\frac{\sqrt{x_K x_0}}{t'}\right)\frac{dt'}{t'}$$
$$= |\nu|\sqrt{KF_0}\int_{\frac{\sqrt{x_K x_0}}{t}}^{\infty} e^{-bs} I_{\pm|\nu|}(s)\frac{ds}{s}, \tag{2.55}$$

where variable of integration t' is changed to $s = \frac{\sqrt{x_K x_0}}{t'}$, and notations $\bar{x}$ and b will be used for

$$\sqrt{x_K x_0} = \bar{x} \qquad \frac{x_K + x_0}{,} 2\sqrt{x_K x_0} = b. \tag{2.56}$$

The order of the Bessel function is $|\nu|$ for absorbing and $(-|\nu|)$ for reflecting case. Below we use Schläfli integral representation of $I_\nu(s)$ [74]

$$I_\nu(s) = \frac{1}{2\pi i}\int_{C_w} e^{s\cosh u - \nu u} du, \tag{2.57}$$

where C_u is the three-legged contour in the complex plane w consisting of two horizontal and one vertical legs,

$$C_u \quad (-i\pi + \infty, -i\pi], \quad [-i\pi, i\pi], \quad [i\pi, i\pi + \infty). \tag{2.58}$$

Integrating by parts, we also have

$$I_\nu(s) = \frac{1}{2\pi i}\frac{s}{\nu}\int_{C_u} e^{s\cosh u - \nu w}\sinh u\, du.$$

Plugging this integral into (2.55), and integrating first with respect to s, we come up with

$$\mathscr{O}(t, K, F_0) = \pm\frac{\sqrt{KF_0}}{2\pi i}\int_{C_u}\frac{e^{-\frac{\bar{x}(b-\cosh\phi)}{t}\mp|\nu|u}}{b-\cosh u}\sinh u\, du.$$

Now expanding the integral over C_u into its component legs we obtain

$$\mathscr{O}(t, K, F_0) = \frac{\sqrt{KF_0}}{\pi}\int_0^\pi \frac{\sinh\phi\sin(|\nu|\phi)}{(b-\cos\phi)} e^{-\frac{\bar{x}(b-\cosh\phi)}{t}} d\phi$$
$$+ \frac{\sqrt{KF_0}}{\pi}\sin(|\nu|\pi)\int_0^\infty \frac{\sinh\psi e^{\mp|\nu|\psi}}{(b+\cosh\psi)} e^{-\frac{\bar{x}(b+\cosh\psi)}{t}} d\psi, \tag{2.59}$$

where we use parametrization $u = i\phi$ on the vertical leg $(-\pi \le \phi \le \pi)$ and $u = \pm i\pi + \psi$ on the horizontal legs $(0 \le \phi)$. Recall also that the upper (lower) sign in the exponential $e^{\mp|\nu|\psi}$ refers to the absorbing (reflecting) solution.

The last expression serves as a base for our further handling of the SABR model.

2.8 CEV Model with Stochastic Volatility

We consider now a more general stochastic volatility (SV) case

$$\mathrm{d}F_t = F_t^{\beta}\sqrt{V_t}\mathrm{d}W_t \quad 0 \le \beta < 1, \tag{2.60}$$

$$\mathrm{d}V_t = \mu_t(V_t)\mathrm{d}t + \phi_t(V_t)\mathrm{d}Z_t, \tag{2.61}$$

$$\mathrm{d}Z_t\mathrm{d}W_t = 0,$$

with the restriction that V_t remains non-negative. We suggest the zero correlation between underlying F and volatility V. Define stochastic time τ, as the cumulative variance

$$\tau = \int_0^t V_{t'}\mathrm{d}t'$$

and the process

$$\mathrm{d}B_\tau = \sqrt{V_t}\mathrm{d}W_t,$$

which is Brownian motion adapted to τ. Then the process F_t becomes the time changed CEV process, its probability density p_{SV} is given by the expectation of the CEV PDF p_{cev} over variance τ [56]

$$p(t, F|F_0) = \mathbb{E}\left[p_{\mathrm{cev}}(\tau, F|F_0)\right]$$

and requires knowledge of the PDF of τ. It follows then, that the time-value $\mathcal{O}$ of an option on F_t, struck at K at an expiry t is the expectation over τ of the CEV option value $\mathcal{O}_{\mathrm{cev}}(\tau, K, F_0)$ [49, 56] (we extend here arguments, originally used in [49] in connection to the log-normal model, to the CEV model)

$$\mathcal{O}_{\mathrm{SV}}(t, K, F_0) = \mathbb{E}\left[\mathcal{O}_{\mathrm{cev}}(\tau, K, F_0)\right].$$

Note, that in particular cases of $\beta = 0, 1$, PDF p_{cev} reduces to the normal, or log-normal PDF.

Using now the CEV expression (2.59) with τ standing for t, we obtain the general result

$$\mathcal{O}_{\mathrm{SV}}(t, K, F_0) = \frac{\sqrt{KF_0}}{\pi}\left(\int_0^{\pi}\frac{\sin(|\nu|\,\phi)\sin\phi}{b-\cos\phi}\mathbb{E}\left[\mathrm{e}^{-\frac{\bar{x}(b-\cos\phi)}{\tau}}\right]\mathrm{d}\phi\right.$$
$$\left.+\sin(|\nu|\,\pi)\int_0^{\infty}\frac{\mathrm{e}^{\mp|\nu|\psi}\sinh\psi}{b+\cosh\psi}\mathbb{E}\left[\mathrm{e}^{-\frac{\bar{x}(b+\cosh\psi)}{\tau}}\right]\mathrm{d}\psi\right). \tag{2.62}$$

We see that the integrand depends on the moment generating function (MGF) of τ^{-1}, namely $\mathbb{E}\left[\exp\left(-\lambda\tau^{-1}\right)\right]$, which is, in general, difficult to calculate. It may be convenient to express it in terms of the MGF of τ, $M_\tau(\lambda) = \mathbb{E}\left[\exp\left(-\lambda\tau\right)\right]$ [28]

$$M_{\tau^{-1}}(\lambda) = 1 - \int_0^\infty M_\tau\left(\frac{z^2}{4\lambda}\right) J_1(z)\,\mathrm{d}z,$$

where $J_1(z)$ is the Bessel function of order one. More generally

$$M_{\tau^{-1}}(\lambda) = \mathrm{e}^{-\frac{\lambda}{\tau_0}} + \int_0^\infty \left(\mathrm{e}^{-\frac{z^2\tau_0}{4\lambda}} - M_\tau\left(\frac{z^2}{4\lambda}\right)\right) J_1(z)\,\mathrm{d}z, \tag{2.63}$$

where $\tau_0 > 0$, below we use $\tau_0 := \mathbb{E}[\tau]$. Details of derivation can be found in [11].

A similar approach was developed by Skabelin in [72], where expansion of MGM of τ was used to calculate the series of option values in terms of variance central moments $\overline{(\tau - \bar{\tau})^n}$. The MGF of τ is known analytically for the class of affine processes V_t, a sub-case of which we consider in Sect. 2.8.1.

The MGF of τ^{-1} is known analytically for the geometric Brownian motion (i.e. SABR) [9, 63], in which case our approach provides a fast 1D integration.

Next, we consider two important applications of our results: the CEV-CIR model (or power generalization of the Heston) and the SABR one.

2.8.1 CEV-CIR

We now specialize to the affine case

$$\mathrm{d}F_t = F_t^\beta \sqrt{V_t}\mathrm{d}W_t \quad 0 \le \beta < 1, \tag{2.64}$$

$$\mathrm{d}V_t = \kappa(\theta - V_t)\mathrm{d}t + \xi\sqrt{V_t}\mathrm{d}Z_t, \tag{2.65}$$

$$\mathrm{d}Z_t \mathrm{d}W_t = 0.$$

If $\xi^2 \le 2\kappa\theta$ and $V_0 > 0$, V_t always stays positive (this Feller [31] condition corresponds to $0 < \nu$ in our BESQ analysis, when $V = 0$ is unattainable). Observe that V_t is a square-root process, and τ is its integral. The MGF of τ, which is known from the study of the CIR model [52], takes the form

$$M_\tau(p) := \mathbb{E}\left[\exp(-p\tau)\right] = \exp(-A - V_0 B), \tag{2.66}$$

$$B = \frac{2p}{\kappa + \gamma\coth(\gamma t/2)},$$

$$A = \frac{2k\theta}{\xi^2}\log\left(1 + \frac{(\gamma+\kappa)\left(e^{(\gamma-\kappa)t/2} - 1\right) + (\gamma-\kappa)\left(e^{-(\gamma+\kappa)t/2} - 1\right)}{2\gamma}\right),$$

$$\gamma = \kappa + \frac{2\xi^2 p}{\sqrt{\kappa^2 + 2\xi^2 p} + \kappa}.$$

We turn then to the general expression (2.62) for the option time value, where the required MGF $M_{\tau^{-1}}$ may be calculated from M_τ using relation (2.63) with

$$\tau_0 := \mathbb{E}[\tau] = (V_0 - \theta)\,\frac{1-\mathrm{e}^{-\kappa T}}{\kappa} + \theta T.$$

More details on calculation of $M_{\tau^{-1}}$, numerical methods applied, and computation of the implied volatility with comparison to the Monte-Carlo simulation can be found in [11].

2.8.2 SABR

Recall the SABR process F_t follows

$$\mathrm{d}F_t = F_t^{\beta} V_t \mathrm{d}W_t, \tag{2.67}$$

$$\mathrm{d}V_t = \gamma V_t \mathrm{d}Z_t. \tag{2.68}$$

Note that we have changed notation here, namely V stands now for $\sqrt{V}$ from (2.60). As a result

$$V_t = V_0 \exp\left(-\frac{1}{2}\gamma^2 t + \gamma Z_t\right),$$

$$\tau = \int_0^t V_{t'}^2 \mathrm{d}t',$$

$$= V_0^2 \int_0^t \exp\left(-\gamma^2 t' + 2\gamma Z_{t'}\right) \mathrm{d}t'. \tag{2.69}$$

It turns out that for the geometric Brownian motion V_t, like in (2.68), it is possible to obtain MGF

$$M_{\tau^{-1}}(\lambda) = \mathbb{E}\left[\exp\left(-\lambda\tau^{-1}\right)\right]$$

in the form of one dimensional integral [63]. We have managed to recast the result [63] into a different form, which is more convenient for option pricing and may be of interest in its own right [11]; the MGF $M_{\tau^{-1}}(\lambda)$ looks like

$$\mathbb{E}\left[\exp\left(-\frac{\lambda}{\tau}\right)\right] = \frac{G(\gamma^2 t, s)}{\cosh s}, \tag{2.70}$$

$$s = \sinh^{-1}\left(\frac{\gamma\sqrt{2\lambda}}{V_0}\right),$$

$$G(t,s) = 2\sqrt{2}\frac{\mathrm{e}^{-\frac{t}{8}}}{t\sqrt{2\pi t}} \int_s^{\infty} \mathrm{d}u\, u\, \mathrm{e}^{-\frac{u^2}{2t}} \sqrt{\cosh u - \cosh s}. \tag{2.71}$$

So defined cumulative distribution function $G(t, s)$ (CDF) has been introduced in [10] along with a geometric interpretation in terms of the Brownian diffusion on the Poincare hyperbolic half-plane $\mathbf{H}^2$. Namely, the CDF $G(t, s)$ represents the probability $P(s(x, y) > s)$ of the hyperbolic distance $s(x, y)$ between observation and source points, with the probability density function of $s(x, y)$ given by the McKean heat kernel $G^{(2)}(t, s)$ (3.15). Derivation of important formulas for MGF (2.70) and CDF (2.71) is given in Chap. 3.

Substituted into (2.62), MGF (2.70) generates the following expression for the option time value

$$\begin{aligned}\mathcal{O}_{\text{SABR}}(t, K, F_0) = \frac{1}{\pi}\sqrt{K F_0}\left\{\int_0^{\pi} \mathrm{d}\phi \, \frac{\sin\phi \sin(|\nu|\,\phi)}{b - \cos\phi} \, \frac{G(\gamma^2 t, s(\phi))}{\cosh s(\phi)}\right. \\ \left. + \sin(|\nu|\,\pi) \int_0^{\infty} \mathrm{d}\psi \, \frac{\sinh\psi}{b + \cosh\psi} \mathrm{e}^{\mp|\nu|\psi} \, \frac{G(\gamma^2 t, s(\psi))}{\cosh s(\psi)}\right\}. \qquad (2.72)\end{aligned}$$

Here s has the following parametrization with respect to ϕ and ψ with $q_K q_0 = \bar{x}$ and b defined in (2.56)

$$\sinh s(\phi) = \frac{\sqrt{2 q_K q_0 (b - \cos\phi)}}{V_0}, \qquad (2.73)$$

$$\sinh s(\psi) = \frac{\sqrt{2 q_K q_0 (b + \cosh\psi)}}{V_0}. \qquad (2.74)$$

2.9 Conclusion

In this chapter, we have considered a CEV model with a general stochastic volatility [11]. Assuming that rate-volatility correlation is zero we have obtained an exact integral representation of the option price provided that we have a closed form for the MGF of the cumulative stochastic variance or of its inverse. Using this result, we have derived explicit solutions for both CEV-CIR (power generalization of the Heston) model and the SABR one.

Chapter 3
Classic SABR Model: Exactly Solvable Cases

3.1 Introduction

In this chapter, we present advanced analytical formulas for SABR model option pricing. We analyze exactly solvable cases, starting with the free Normal SABR and Log-normal SABR, both being of the practical and methodological interest. We consider then the Normal SABR (positive rate) with zero correlation, and a general zero correlation case. We use results for the normal free SABR and for the zero correlation case to construct the mixed SABR model in the Chap. 5.

3.2 Probability Density Functions for Free Normal and Log-Normal SABR, Probabilistic Approach

Here we obtain joint distributions of underlying F and volatility V for the normal ($\beta = 0$) and the log-normal ($\beta = 1$) SABR model using the probabilistic approach. We pass to the process $Q = \int F^{-\beta} \mathrm{d}F$, which is $Q_t = F_t$ for $\beta = 0$ and $Q_t = \log F_t$ for $\beta = 1$. Together with Q, V consider cumulative variance $I_t = \int_0^t V_{t'}^2 \mathrm{d}t'$, which will serve as a stretched random time τ (termed subordinator in the theory of stochastic processes). Driving SDEs for processes Q, V, I are

$$\mathrm{d}Q_t = V_t \, \mathrm{d}W_t^1 - \frac{\beta}{2} V_t^2 \mathrm{d}t\,, \tag{3.1}$$

$$\mathrm{d}V_t = V_t \, \mathrm{d}W_t\,, \tag{3.2}$$

$$\mathrm{d}I_t = V_t^2 \mathrm{d}t\,, \tag{3.3}$$

recall also, we assume here the zero correlation between W^1 and W.

Probability distribution of Q_t and V_t is obtained as a marginal distribution by integrating random time τ off the triple joint PDF of Q_t, V_t and I_t (for brevity we omit dependence on real time t)

A. Antonov et al., *Modern SABR Analytics*,
SpringerBriefs in Quantitative Finance,
https://doi.org/10.1007/978-3-030-10656-0_3

$$p(q,V) = \int_0^\infty p(q,V,\tau)\mathrm{d}\tau \,.$$

In turn, the triple joint PDF is expressed through the conditional PDF of Q and V given τ and the probability density of random time τ

$$p(q,V,\tau) = p(q,V\,|\tau)\;p(\tau)\,.$$

At zero correlation, PDF $p(q,V\,|\tau)$ is factorized into the product of conditional distributions of q given τ and of V given τ,

$$p(q,V\,|\tau) = p(q\,|\tau)\;p(V\,|\tau) \tag{3.4}$$

Collecting pieces we get

$$p(q,V,\tau) = p(q\,|\tau)\;p(V\,|\tau)\;p(\tau) = p(q\,|\tau)\;p(V,\tau)\,, \tag{3.5}$$

where $p(V,\tau)$ is joint PDF of V and τ. While other steps follow regular relations between conditional and absolute probabilities, factorization (3.4) (even if quite intuitive for uncorrelated processes Q_t and V_t) requires in fact a stronger property, namely that Q_t and V_t be independent, e.g., not linked by any global condition. This is true in case of zero correlation, $\rho = 0$, as boundary conditions on the probability density are imposed autonomously for Q_t and for V_t. Consider, though, SABR model with $\rho \neq 0$ (and $\bar\rho = \sqrt{1-\rho^2}$). We can exclude local correlation by decomposing the first Brownian motion as $W_t^1 = \rho\,W_t + \bar\rho\;Z_t$, with $\mathrm{d}Z_t$ not correlated with $\mathrm{d}W_t$, and mixing processes $Q_t = F_t$ and V_t into the new ones $X_t = (F_t - \rho V_t)/\bar\rho$ and $Y_t = V_t$. The new driving SDEs, $\mathrm{d}X_t = Y_t\;\mathrm{d}Z_t$ and $\mathrm{d}Y_t = Y_t\;\mathrm{d}W_t$ show that increments of processes X_t and Y_t are uncorrelated. However, if we require that underlying F_t be positive and impose the appropriate boundary condition (absorbing or reflecting) at $F = 0$, it translates into the boundary condition on the straight line $\bar\rho\;x + \rho\;y = 0$, thus causing a global dependence between X_t and Y_t and destroying factorization of $p(x,y\,|\tau)$.[1]

Returning to our case, we observe that joint PDF $p(V,\tau)$ in (3.5) is related to the joint distribution of a geometric Brownian motion and an accumulated integral of its exponential, obtained by Yor [63, 75] and adapted in [9] to the form

$$p(V,\tau) = \frac{\mathrm{e}^{-t/8}}{V^2\,\tau}\,(V\,V_0)^{1/2}\,\mathrm{e}^{-\frac{V^2+V_0^2}{2\tau}}\;\theta\left(\tfrac{V\,V_0}{\tau},t\right)\,, \tag{3.6}$$

$$\theta(r,t) = \frac{1}{\mathrm{i}\,(2\pi t)^{3/2}}\int_{C_u} \mathrm{e}^{r\cosh u - u^2/2t}\,u\,\mathrm{d}u\,, \tag{3.7}$$

[1] We have found the solution for the joint PDF of F and V for the normal SABR with $\rho \neq 0$ and $F \geq 0$ by quadratures, but the expression is too complicated to be of a practical use.

where the path of integration C_u is defined in (2.58), recall, it consists of three sides of the infinite rectangle with corners at $+\infty - \mathrm{i}\pi, \quad -\mathrm{i}\pi, \quad +\mathrm{i}\pi, \quad +\infty + \mathrm{i}\pi$. Despite expression (3.7) looks different from the original one [75]

$$\theta_Y(r,t) = \frac{r}{\sqrt{2\pi^3 t}} \int_0^\infty \mathrm{e}^{-r\cosh\xi + (\pi^2 - \xi^2)/2t} \sinh\xi \, \sin(\pi\xi/t)\, \mathrm{d}\xi \,, \tag{3.8}$$

they, in fact, describe the same function. To show it, consider three legs of contour C_u. The vertical leg, $(-\mathrm{i}\pi, \mathrm{i}\pi)$ makes no contribution, because the integrand is an odd function of u. Next, on the upper horizontal leg $(\mathrm{i}\pi, \infty + \mathrm{i}\pi)$ make substitution $u = \xi + \mathrm{i}\pi$ with real ξ running from 0 to $+\infty$ while on the lower leg $(\infty - \mathrm{i}\pi, -\mathrm{i}\pi)$ use $u = -\xi - \mathrm{i}\pi$ with ξ running from $-\infty$ to 0, thus recasting (3.7) into

$$\theta(r,t) = \frac{1}{\mathrm{i}\,(2\pi t)^{3/2}} \int_{-\infty}^{\infty} \mathrm{e}^{-r\cosh\xi - (\xi + \mathrm{i}\pi)^2/2t} \, (\xi + i\pi)\, \mathrm{d}\xi \,.$$

The last expression is transformed into (3.8) after integrating by parts, developing the exponential with $(\xi + \mathrm{i}\pi)^2$, and keeping only the even part of the integrand. Even though representation (3.8) is explicitly real, we prefer to keep $\theta(r,t)$ in the form of the contour integral (3.7), which is more efficient in utilizing analytic properties of functions in the complex plane u.

Turning to conditional distribution $p(q\,|\tau)$, we recall that after adopting cumulative variance I_t as a random time τ, process $\mathrm{d}B_\tau = V_t \mathrm{d}W_t^1$ becomes τ-time Brownian motion, process Q_t (3.1) is also transformed into τ-Brownian motion with ($\beta = 1$) or without drift ($\beta = 0$),

$$\mathrm{d}Q = \mathrm{d}B_\tau - \tfrac{\beta}{2}\mathrm{d}\tau \,.$$

Next we consider separately the normal and log-normal cases.

3.2.1 *Normal SABR*

If F is allowed to take any real values, process $Q = F$ is τ- Brownian motion with the standard Gaussian distribution

$$p_G(F\,|\tau) = \frac{1}{\sqrt{2\pi\tau}} e^{-\frac{(F-F_0)^2}{2\tau}} \,. \tag{3.9}$$

If rate F is restricted to positive values, then either absorbing (a), or reflecting (r) condition must be imposed at $F = 0$, accordingly, the distributions become

$$p_{a/r}(F\,|\tau) = \frac{1}{\sqrt{2\pi\tau}} \mathrm{e}^{-\frac{(F-F_0)^2}{2\tau}} \mp \frac{1}{\sqrt{2\pi\tau}} \mathrm{e}^{-\frac{(F+F_0)^2}{2\tau}} \tag{3.10}$$

Till the very end, we will keep notations q, q_0 for F, F_0 in order to reuse the same expressions in the next case of $\beta = 1$.

Whether dealing with the standard Gaussian or absorbing (reflecting) distribution, it suffices to compute the Gaussian component. Combining (3.6, 3.9), we get

$$\begin{aligned} p(q, V) &= \int_0^\infty p_G(q\,|\tau)\; p(V, \tau)\, d\tau \\ &= \frac{\mathrm{e}^{-t/8}}{V^2}\left(\frac{V V_0}{2\pi}\right)^{1/2} \int_0^\infty \tau^{-3/2}\, \mathrm{e}^{-\frac{(q-q_0)^2+V^2+V_0^2}{2\tau}}\, \theta(\tfrac{V V_0}{\tau}, t)\, \mathrm{d}\tau\,. \end{aligned} \tag{3.11}$$

It is at this point that the geodesic distance $s(q, V; q_0, V_0)$ on the hyperbolic half-plane $\mathbf{H}^2$ appears on the scene,

$$\cosh s = \frac{(q-q_0)^2 + V^2 + V_0^2}{2V V_0}\,. \tag{3.12}$$

So defined, $\cosh s \geq 1$ with minimum $s = 0$ at $q = q_0,\ V = V_0$.

Next, we introduce the invariant probability density $p_{\mathrm{inv}} = V^2 p(q, V)$, which is PDF with respect to the invariant, or Riemannian element of volume $\mathrm{d}V_{\mathrm{inv}} = \frac{\mathrm{d}q\mathrm{d}V}{V^2}$ in $\mathbf{H}^2$, then we plug $\cosh s$ into (3.11), and change variable of integration to $r = \frac{V V_0}{\tau}$, coming up with

$$p_{\mathrm{inv}} = \frac{\mathrm{e}^{-t/8}}{\sqrt{2\pi}} \int_0^\infty r^{-1/2}\, \mathrm{e}^{-r\cosh s}\, \theta(r, t)\, \mathrm{d}r\,. \tag{3.13}$$

Observe that PDF p_{inv} depends only on time t and distance s (3.12). Being the true scalar, p_{inv} may be more convenient to use, because it keeps its value under any change of variables and needs only to be expressed in terms of new ones (with no Jacobian of transition involved).

Now substitute function $\theta(r, t)$ (3.7) into the last expression for p_{inv} and integrate over r first,

$$\int_0^\infty \mathrm{e}^{-Ar}\, r^{-1/2}\, \mathrm{d}r = \left(\frac{\pi}{A}\right)^{1/2} \qquad A = \cosh s - \cosh u\,.$$

Then PDF p_{inv} (3.13) takes the form

$$p_{\mathrm{inv}} = \frac{\sqrt{2}\,\mathrm{e}^{-t/8}}{(2\pi t)^{3/2}} \frac{1}{2\mathrm{i}} \int_{C_u} \frac{u\, \mathrm{e}^{-u^2/2t}}{(\cosh s - \cosh u)^{1/2}}\, \mathrm{d}u\,. \tag{3.14}$$

Notice that the above factor A is positive at all complex u on contour C_u (with positive value of $A^{1/2}$ taken), because on horizontal legs $\cosh u = \cosh(\Re u \pm \mathrm{i}\pi) = -\cosh \Re u < 0$, and on the vertical leg, $u = \mathrm{i}\phi$, $\cosh u = \cos\phi \leq 1 < \cosh s$.

As the last step, we take advantage of presenting PDF p_{inv} in the form of the contour integral (3.14). We consider analytic properties of the integrand as a function of the complex variable u and deform contour C_u as explained in Appendix 3.8. The main formula (3.133) reads now as

$$\frac{1}{2\mathrm{i}}\int_{C_u}\frac{u\mathrm{e}^{-u^2/2t}}{(\cosh s-\cosh u)^{1/2}}\,du=\int_s^\infty\frac{u\,\mathrm{e}^{-u^2/2t}}{(\cosh u-\cosh s)^{1/2}}\,du\,.$$

Using this transformation in (3.14) immediately leads to the classical McKean formula [62]

$$p_{\mathrm{inv}}(t,s):=G^{(2)}(t,s)=\frac{\sqrt{2}\mathrm{e}^{-t/8}}{(2\pi t)^{3/2}}\int_s^\infty\frac{u\,\mathrm{e}^{-u^2/2t}}{(\cosh u-\cosh s)^{1/2}}\,du\,, \tag{3.15}$$

which is the invariant probability density for Brownian diffusion (a.k.a. the heat kernel) on the hyperbolic half-plane $\mathbf{H}^2$ (recall that $q=F$ for $\beta=0$).

For the normal SABR model per se, rate F must be positive, with a boundary condition imposed at $F=0$. In the realistic case of the absorbing conditional PDF $p_a(F\mid\tau)$ (3.10) we come up with the joint probability density expressed as the difference of direct and reflected McKean kernels,

$$p_{\mathrm{inv}}^{(a)}(F,V)=G^{(2)}(t,s_d)-G^{(2)}(t,s_r), \tag{3.16}$$

where distances s_d and s_r in $\mathbf{H}^2$ are defined as

$$\cosh s_{d/r}=\frac{(F\mp F_0)^2+V^2+V_0^2}{2VV_0} \tag{3.17}$$

Remind, this solution works only for zero correlation, $\rho=0$.

For the free normal SABR with $\rho\neq 0$, $p_{\mathrm{inv}}(t,s)$ keeps its form (3.15) with distance s expressed through original variables, as defined below (3.38).

3.2.1.1 Other Derivations of the McKean Heat Kernel

We refer to the three ones,[2] all involving special functions and their integral representations. Hagan et al. [42] used the spectral expansion of the sought p_{inv} in terms of Legendre functions of the first kind and applied related Meller-Fock transformations to obtain the result (3.15). Book [57] (Lewis) and review [63] (Matsumoto-Yor, MY) used the fact that the Laplace transform of kernel (3.15) is given by Legendre function of the second kind $Q_\nu(\cosh s)$ with known integral representation [34], 8.715. Lewis computed the double Laplace-Fourier transform of the solution, then identified the inverse Fourier transform as function $Q_\nu(\cosh s)$ [34], 6.672. MY [63] applied the probabilistic approach to the problem of Brownian diffusion in $\mathbf{H}^n$ and came up with

[2]It is interesting, that McKean paper is not in this list. Formula (3.15) appeared in [62] as a side result, maybe too obvious, or well known, without any derivation or reference thereof.

the new formula [35] for p_{inv} in dimension $n = 2$

$$p_2 = \frac{e^{-t/8}}{4(\pi^3 t)^{1/2}} \int_0^\infty \frac{e^{-\xi^2/2t} \sinh\xi \; \sin(\pi\xi/t)}{(\cosh s + \cosh\xi)^{3/2}} \, d\xi \,, \tag{3.18}$$

which is identical to our intermediate expression (3.14) (integrate (3.14) by parts and develop legs of C_u like earlier for function θ). MY then computed Laplace transform of the expression, preceding this one (analog of our (3.13)) and identified the remaining integral over r as the proper Legendre function.

All these derivations (ours not excluded) are rather involved.

3.2.2 Log-Normal SABR, $\beta = 1$

Process $Q = \log F$ is obviously not restricted by sign implying that unlike the previous case ($\beta = 0$), generalization for nonzero correlation is quite straightforward for $\beta = 1$. Making the same decomposition $dW_1 = \rho dW + \bar{\rho} dZ$ with $\mathbb{E}[dWdZ] = 0$ and introducing processes

$$X = \frac{Q - \rho V}{\bar{\rho}}, \qquad Y = V, \tag{3.19}$$

we get driving SDEs

$$dX = Y dZ - \tfrac{1}{2\bar{\rho}} Y^2 dt, \qquad dY = Y dW.$$

Further scaling $\tilde{X} = X/\bar{\rho}$ and $\tilde{Y} = Y/\bar{\rho}$ allows also to exclude $\bar{\rho}$ from drift and get SDEs that literally coincide with original ones for Q_t and V_t (3.1, 3.2). New processes X_t and Y_t (or $\tilde{X}$ and $\tilde{Y}$) are uncorrelated and are globally independent because boundary conditions in terms of X_t and Y_t remain autonomously imposed at $x = \pm\infty$ and $y = 0, \infty$, so that factorization of probability distribution $p(x, y\,|\tau)$ does take place like in (3.4). We proceed with zero correlation, but should we turn to the nonzero case we will have to merely substitute $\tilde{x}$ and $\tilde{y}$ for q and V in the final expression $p_{\text{inv}}(q, V)$.

As discussed above, process Q, adapted to random time τ, becomes a standard (not stopped) Brownian motion with drift,

$$dQ = dB_\tau - \tfrac{1}{2} d\tau \,,$$

its probability density conditioned on given τ looks like

$$p(q\,|\tau) = \frac{1}{\sqrt{2\pi\tau}} e^{-\frac{(q-q_0+\tau/2)^2}{2\tau}} = \frac{1}{\sqrt{2\pi\tau}} e^{-\frac{(q-q_0)^2}{2\tau} - \frac{(q-q_0)}{2} - \frac{\tau}{8}} \,.$$

Plugging it into analog of (3.11) we get

$$
\begin{aligned}
p_{\text{inv}}(q, V) &= V^2 \int_0^\infty p(q\,|\tau)\, p(V, \tau)\, d\tau \\
&= \mathrm{e}^{-t/8-(q-q_0)/2} \left(\frac{V V_0}{2\pi}\right)^{1/2} \int_0^\infty \tau^{-3/2}\, \mathrm{e}^{-\frac{\tau}{8}-\frac{(q-q_0)^2+V^2+V_0^2}{2\tau}}\, \theta\left(\tfrac{V V_0}{\tau}, t\right) \mathrm{d}\tau\,.
\end{aligned}
\tag{3.20}
$$

The last integrand differs from its analog (3.11) only by factor $\mathrm{e}^{-\tau/8}$, but the result, of course, is very different. Again, use expression (3.7) for function $\theta(r, t)$ and consider integral over τ, which is of the type

$$
\frac{b}{\sqrt{2\pi}} \int_0^\infty \mathrm{e}^{-\frac{a^2}{2\tau}-\frac{b^2\tau}{2}}\, \tau^{-3/2}\, \mathrm{d}\tau = \mathrm{e}^{-ab}\,,
\tag{3.21}
$$

studied later more generally, for finite limits $(0, T)$ (3.98, 3.99). Using

$$
a^2 = 1/4, \quad b^2 = 2 V V_0 \left(\cosh s - \cosh u\right),
$$

we find

$$
p_{\text{inv}}(q, V) = \frac{\sqrt{2}\mathrm{e}^{-\frac{t}{8}-\frac{q-q_0}{2}}}{(2\pi t)^{3/2}} \frac{1}{2\mathrm{i}} \int_{C_u} \frac{u\, \mathrm{e}^{-\frac{u^2}{2t}}}{(\cosh s - \cosh u)^{1/2}} e^{-\sqrt{(V V_0(\cosh s - \cosh u)/2}}\, \mathrm{d}u\,.
$$

Here we again invoke analysis of analytic properties of the integrand from Appendix 3.8. Using now formula (3.134) we obtain the 'log-normal' invariant PDF

$$
\begin{aligned}
p_{\text{inv}}(q, V) &= V^2\, p(q, V) \\
&= \frac{\sqrt{2}\mathrm{e}^{-\frac{t}{8}-\frac{q-q_0}{2}}}{(2\pi t)^{3/2}} \int_s^\infty \frac{u\, \mathrm{e}^{-\frac{u^2}{2t}}}{\sqrt{\cosh u - \cosh s}} \cos\left(\frac{\sqrt{2 V V_0}\sqrt{\cosh u - \cosh s}}{2}\right) \mathrm{d}u\,.
\end{aligned}
\tag{3.22}
$$

It can be rewritten also in a more compact form

$$
p_{\text{inv}}(q, V) = \frac{e^{-\frac{q-q_0}{2}-\frac{t}{8}}}{\sqrt{V V_0}(2\pi t)^{3/2}} 2 \int_0^\infty \frac{u\, e^{-\frac{u^2}{2t}}}{\sinh u} \cos\frac{q_2}{2} \mathrm{d}q_2
\tag{3.23}
$$

by introducing new variable of integration q_2 instead of u,

$$
q_2 = \sqrt{2 V V_0}\sqrt{\cosh u - \cosh s}\,.
$$

Parameter u becomes now function of q, q_2 and V,

$$\cosh u = \cosh s + \frac{q_2^2}{2VV_0} = \frac{(q-q_0)^2 + q_2^2 + V^2 + V_0^2}{2VV_0}, \tag{3.24}$$

thus clearly indicating relation to the geometry of the hyperbolic space $\mathbf{H}^3$ with 'vertical' coordinate V and two 'horizontal' coordinates q, q_2 (this explains notation q_2 as the second horizontal coordinate). Namely u represents the hyperbolic distance between points (q, q_2, V) and $(q_0, 0, V_0)$. Moreover, the integrand in (3.23) is associated with the invariant heat kernel in $\mathbf{H}^3$, as discussed in the next Section.

To generalize for **non-zero correlation**, it is sufficient to make in p_{inv} (3.23) all previous substitutions in reverse,

$$q - q_0 \rightarrow \tilde{x} - \tilde{x}_0 = \frac{x - x_0}{\bar{\rho}} = \frac{q - q_0 - \rho(V - V_0)}{\bar{\rho}^2},$$

$$q_2 \rightarrow \tilde{x}_2 = \frac{x_2}{\bar{\rho}}, \qquad V, V_0 \rightarrow \tilde{y}, \tilde{y}_0 = \frac{y, y_0}{\bar{\rho}} = \frac{V, V_0}{\bar{\rho}},$$

with hyperbolic distance u defined now as

$$\cosh u = \frac{\frac{1}{\bar{\rho}^2}[q - q_0 - \rho(V - V_0)]^2 + x_2^2 + V^2 + V_0^2}{2VV_0}, \tag{3.25}$$

Recall that $q - q_0 = \log \frac{F}{F_0}$ and $\bar{\rho}^2 = 1 - \rho^2$. It should be noted also that the element of probability is $\mathrm{d}P = p_{\text{inv}}\mathrm{d}V_{\text{inv}}$ with the Riemannian volume element $\mathrm{d}V_{\text{inv}}$ taking different forms, as needed

$$\mathrm{d}V_{\text{inv}} = \frac{\mathrm{d}\tilde{x}\mathrm{d}\tilde{y}}{\tilde{y}^2} = \frac{\mathrm{d}x\mathrm{d}y}{y^2} = \frac{\mathrm{d}q\mathrm{d}V}{\bar{\rho}V^2} = \frac{\mathrm{d}F\mathrm{d}V}{\bar{\rho}FV^2}.$$

3.3 Deriving PDFs Using Kolmogorov Equations

We have derived the normal and log-normal SABR probability densities using probabilistic arguments. Now we follow up how these PDFs arise as solutions of Kolmogorov equations related in turn to heat equations in hyperbolic spaces. Analysis of the Kolmogorov equation (KE) becomes more consistent and lucid if the state space is endowed with metric that allows introducing distances, invariant volume, and other attributes of a Riemannian manifold. The classical result of Varadhan [76]

$$\lim_{t \to 0} t \, \log G_{x_0}(t; x) = -\frac{s(x, x_0)^2}{2}$$

links the leading order of the short time asymptotics of the KE Green function $G(t, x \,|x_0)$ to the geodesic distance s in a Riemannian metric which is determined by the coefficients of the Kolmogorov equation. Namely, the upper (contravariant) metric tensor g^{ij} is identified as the covariance matrix of the diffusion operator, $\hat{L}_2 = g^{ij}\partial_i\partial_j$. Then the lower (covariant) metric tensor g_{ij} which is inverse to g^{ij}, determines the element of distance $\delta l^2 = g_{ij}\mathrm{d}x^i\mathrm{d}x^j$, and the metric determinant $g = \det(g_{ij})$ serves to define the Riemannian volume element $\mathrm{d}V_{\mathrm{inv}} = \sqrt{g}\,\Pi\mathrm{d}x^i$ and its natural partner – the invariant delta function (both are scalars) $\delta_{\mathrm{inv}}(z, z_0) = g^{-1/2}\delta(z - z_0)$

$$\delta_{\mathrm{inv}}(z, z_0) = g^{-1/2}\delta(z - z_0)\,.$$

Like earlier, we start with $Q = \int F^{-\beta}dF$, which are $Q_t = F_t$ for $\beta = 0$ and $Q_t = \log F_t$ for $\beta = 1$. The driving SDEs are

$$\mathrm{d}Q_t = V_t\,\mathrm{d}W_t^1 - \frac{1}{2}\beta\,V_t^2\mathrm{d}t \tag{3.26}$$

$$\mathrm{d}V_t = V_t\,\mathrm{d}W_t \tag{3.27}$$

with $\mathbb{E}[\mathrm{d}W_t^1\mathrm{d}W_t] = \rho\mathrm{d}t$. The joint probability density $p(q, V)$ solves the forward Kolmogorov equation,

$$\partial_t p = \frac{1}{2}\left(\beta\partial_q + \partial_q^2 + 2\rho\partial_q\partial_V + \partial_V^2\right)V^2 p \tag{3.28}$$

thus defining the upper and lower metric tensors and metric determinant as

$$g^{ij} = V^2\begin{pmatrix}1 & \rho\\ \rho & 1\end{pmatrix},\; g_{ij} = (\bar{\rho}V)^{-2}\begin{pmatrix}1 & -\rho\\ -\rho & 1\end{pmatrix},\; g = (\bar{\rho}V^2)^{-2}$$

The Riemannian, or invariant, volume element and the invariant delta function look like

$$\mathrm{d}V_{\mathrm{inv}} = \frac{\mathrm{d}q\mathrm{d}V}{\bar{\rho}\,V^2},\quad \delta_{\mathrm{inv}}(z, z_0) = \bar{\rho}V^2\delta(q - q_0)\delta(V - V_0) \tag{3.29}$$

The invariant probability density (with respect to $\mathrm{d}V_{\mathrm{inv}}$) is defined as $p_{\mathrm{inv}} = \bar{\rho}V^2\,p(q, V)$ and satisfies the heat equation

$$\partial_t p_{\mathrm{inv}} = \frac{V^2}{2}\left(\beta\partial_q + \partial_q^2 + 2\rho\partial_q\partial_V + \partial_V^2\right)p_{\mathrm{inv}} = \frac{1}{2}\left(\beta V^2\partial_q + \Delta\right)p_{\mathrm{inv}} \tag{3.30}$$

where

$$\Delta = \frac{1}{\sqrt{g}}\partial_i\sqrt{g}g^{ij}\partial_j \tag{3.31}$$

is the Laplace-Beltrami operator in the metric just defined. The initial distribution of p_{inv} is $\delta_{\mathrm{inv}}(z, z_0)$ (3.29). As already mentioned, using the invariant PDF has that advantage that being a true scalar, p_{inv} sustains its value under any change of variables and only needs to be expressed in terms of new ones.

Using the 'diagonalizing' transform

$$x = \frac{q - \rho V}{\sqrt{1-\rho^2}}, \qquad y = V \tag{3.32}$$

and passing to derivatives with respect to new coordinates x, y recast the Laplacian and other geometric elements into those on the Poincare half-plane $\mathbf{H}^2$,

$$\delta l^2 = g_{ij} \mathrm{d}x^i \mathrm{d}x^j = \frac{\mathrm{d}x^2 + \mathrm{d}y^2}{y^2}, \qquad \Delta^{(2)} = y^2(\partial_x^2 + \partial_y^2), \tag{3.33}$$

$$\mathrm{d}V_{\mathrm{inv}} = \frac{\mathrm{d}x\mathrm{d}y}{y^2}, \quad \delta_{\mathrm{inv}} = y_0^2\, \delta_x\, \delta_y\,. \tag{3.34}$$

PDF itself solves the equation

$$\partial_t p_{\mathrm{inv}} = \frac{1}{2}\left(\frac{\beta}{\bar{\rho}} y^2 \partial_x + \Delta^{(2)}\right) p_{\mathrm{inv}} \tag{3.35}$$

From now on we will follow cases $\beta = 0$ and $\beta = 1$ separately.

3.3.1 The Normal SABR, ($\beta = 0$)

Invariant PDF p_{inv} solves the canonical heat equation[3] in $\mathbf{H}^2$,

$$\partial_t p_{\mathrm{inv}} = \frac{1}{2} y^2(\partial_x^2 + \partial_y^2) p_{\mathrm{inv}} \tag{3.36}$$

with initial condition $p_{\mathrm{inv}}|_{t=0} = \delta^{(2)}_{\mathrm{inv}}$.

For the **free boundary** Normal SABR, diffusion spreads across the whole half-plane $y \geq 0$, and the solution to both the original (3.30) and the diagonalized (3.36) Kolmogorov equation is given by the McKean heat kernel $G^2(t, s)$ (3.15), obtained above with the probabilistic approach. In Sect. 3.4 $G^2(t, s)$ is derived by solving heat equation (3.36). Distance s is defined through

$$\cosh s = \frac{(x - x_0)^2 + y^2 + y_0^2}{2yy_0} \tag{3.37}$$

in orthogonal coordinates (x, y) (3.32), or by

$$\cosh s = \frac{1}{2VV_0}\left(\frac{1}{\bar{\rho}^2}[F - F_0 - \rho(V - V_0)]^2 + V^2 + V_0^2\right) \tag{3.38}$$

[3]It is due to specifics of the normal SABR model and to space dimension (2) that the diffusion operator in the Kolmogorov equation coincides with Laplacian. In a general case, including log-normal SABR, it contains also additional terms, connection and charge [43, 67].

in original ones (F, V) (recall that now $q = F$).

In the case of zero correlation, with only positive rates F allowed, the invariant PDF (absorbing) is represented by difference of the direct and reflected kernels $G^{(2)}$, as already discussed in Sect. 3.2, (3.16). Each kernel depends on its own distance, s_d (direct), or s_r (reflected) (3.17), which are distances from the observation point (F, V) to the source (F_0, V_0), or to the reflected source $(-F_0, V_0)$). From the standpoint of Kolmogorov equation (3.28) (with $\beta = 0$ and $q = F$), this construction is based in case $\rho = 0$ on both reflection and shift symmetries with respect to F. If $\rho \neq 0$, the reflection symmetry is being lost due to presence of the mixed derivative $\partial_F \partial_V$.

Finally, we present the important relation between PDF $G^{(2)}(t, s)$ and the associated cumulative distribution function (CDF) $G(t, s)$, that has been introduced in [10] and defined in Sect. 3.4,

$$G(t, s) = 2\pi \int_s^\infty G^{(2)}(t, s') \sinh s' \mathrm{d}s', \tag{3.39}$$

its explicit expression is given by (3.82).

Since invariant PDF $G^{(2)}(t, s)$ depends only on t and s, it solves (3.36) with the 'radial' Laplacian (3.62) ($n = 2$)

$$\partial_t G^{(2)}(t, s) = \frac{1}{2}(\sinh s)^{-1} \partial_s \left(\sinh s \, \partial_s G^{(2)}(t, s)\right) . \tag{3.40}$$

Integrating this equation with weight $2\pi \sinh s$ and using definition (3.39) gives the relation

$$\partial_t G(t, s) = -\pi \sinh s \, \partial_s G^{(2)}(t, s) , \tag{3.41}$$

which is instrumental in deriving option values.

Since PDF $G^{(2)}(t, s)$ and CDF $G(t, s)$ are actively used throughout the book, it is appropriate to provide an additional derivation of $G^{(2)}$. In the next Section we present a new and quite simple one, utilizing some basic properties of hyperbolic geometry.

3.3.2 *The Log-Normal SABR* ($\beta = 1$)

In this case the Kolmogorov equation (3.35) contains a drift

$$\partial_t p_{\text{inv}} = \frac{y^2}{2} \left(\frac{1}{\bar{\rho}} \partial_x + \partial_x^2 + \partial_y^2 \right) p_{\text{inv}}$$

that gives rise to connection and charge terms [43, 67] after completing the square with respect to x-derivative,

$$\partial_t p_{\text{inv}} = \frac{y^2}{2}\left((\partial_x + a)^2 + \partial_y^2 - a^2\right) p_{\text{inv}}\,, \tag{3.42}$$

here notation $a = \frac{1}{2\bar{\rho}}$ is used. Constant connection a is excluded by substitution

$$p_{\text{inv}} = \tilde{p}\mathrm{e}^{-a(x-x_0)}$$

resulting in

$$\partial_t \tilde{p} = \frac{1}{2}y^2\left(\partial_x^2 + \partial_y^2 - a^2\right)\tilde{p} \tag{3.43}$$

Initial condition for $\tilde{p}$ is the same as for p_{inv}

$$\tilde{p}_{|t=0} = y_0^2\delta(x - x_0)\delta(y - y_0) \tag{3.44}$$

The difficulty with Eq. (3.43) is that charge $Q = \frac{1}{2}y^2a^2$ is not constant and cannot be excluded. As Henry-Labordère (HL) noticed [43], in order to solve (3.43), it is productive to enter a higher dimension space. Introduce new variable x_2 (also denote temporarily $x - x_0 = x_1$ for the sake of symmetry), then treat $\tilde{p}(t, y, x_1)$ as the Fourier transform of a new function $G^{(2+1)}(t, y, x_1, x_2)$ with respect to x_2 with wave number $k = a$

$$\tilde{p}(\cdot) = \int G^{(2+1)}(\cdot, x_2)\mathrm{e}^{-\mathrm{i}ax_2}\mathrm{d}x_2 \tag{3.45}$$

As known, the second derivative, subject to the Fourier transform, becomes the algebraic factor, $\partial_x^2 \to -k^2$. Accordingly, term $-a^2\tilde{p}$ in (3.43) becomes the Fourier transform of $\partial_{x_2}^2 G^{(2+1)}$, and the whole equation (3.43) becomes the Fourier transform of the following diffusion equation

$$\partial_t G^{(2+1)} = \frac{1}{2}y^2\left(\partial_{x_1}^2 + \partial_{x_2}^2 + \partial_y^2\right)G^{(2+1)} \tag{3.46}$$

Presence of coefficients y^2 at second derivatives clearly points to the geometry of the hyperbolic space $\mathbf{H}^3$ (with element of length $\delta l^2 = y^{-2}(\mathrm{d}x_1^2 + \mathrm{d}x_2^2 + \mathrm{d}y^2)$. However, Eq. (3.46) is not the canonical heat equation in $\mathbf{H}^3$,

$$\partial_t G = \frac{1}{2}\Delta^{(3)}G\,, \tag{3.47}$$

as the latter would involve the Laplace-Beltrami operator $\Delta^{(3)}$ (3.59) ($n = 2$)

$$\Delta^{(3)} = y^2\left(\partial_{x_1}^2 + \partial_{x_2}^2 + \partial_y^2 - y^{-1}\partial_y\right) \tag{3.48}$$

that contains additional term $(-y\partial_y)$ against the diffusion operator in (3.46).

Fortunately, diffusion equation (3.46) can be transformed into heat equation (3.47), as described in the next Section, and its solution $G^{(2+1)}$ (3.74) can be expressed through the fundamental solution $G^{(3)}$ (3.70) of Eq. (3.47),

$$
\begin{aligned}
G^{(2+1)}(t, x - x_0, x_2, y) &= \frac{e^{\frac{3}{8}t}}{\sqrt{y y_0}} G^{(3)}(t, u)\,, \\
G^{(3)}(t, u) &= (2\pi t)^{-3/2}\, e^{-\frac{t}{2} - \frac{u^2}{2t}}\, \frac{u}{\sinh u}\,, \\
\cosh u(x, x_2, y;\, x_0, 0, y_0) &= \frac{(x - x_0)^2 + x_2^2 + y^2 + y_0^2}{2 y y_0}\,,
\end{aligned}
\tag{3.49}
$$

The probability density p_{inv} is now obtained by taking Fourier transform (3.45), thus reproducing the earlier result [73] (3.23),

$$
p_{\mathrm{inv}} = \frac{e^{-\frac{x - x_0}{2\bar{\rho}} - \frac{t}{8}}}{\sqrt{y y_0}\,(2\pi t)^{3/2}}\, 2 \int_0^{\infty} e^{-\frac{u^2}{2t}}\, \frac{u}{\sinh u} \cos\frac{x_2}{2\bar{\rho}}\, \mathrm{d}x_2\,, \tag{3.50}
$$

Expression (3.49) defines distance u in terms of orthogonal coordinates (x, y) (3.32) and turns into earlier one (3.25) in terms of original variables (q, V) $(q - q_0 = \log(F/F_0)$. Recall also that the probability density of variables F and V is related to the invariant PDF as

$$
p(F, V) = \sqrt{g(F, V)}\, p_{\mathrm{inv}} = \frac{p_{\mathrm{inv}}}{\bar{\rho} F V^2}\,. \tag{3.51}
$$

A similar solution to the log-normal SABR is obtained in the book by Lewis [57] as a result of series of variable transformations in the Kolmogorov equation. On the other hand, solution claimed in [43], is incorrect (despite the insightful suggestion to add an auxiliary variable x_2). The thing is that the diffusion operator in (3.46) was confused in [43] with the Laplacian $\Delta^{(3)}$ (3.48), and solution $G^{(3)}$ to the canonical heat equation in $\mathbf{H}^3$ was mistakenly used in the capacity of $G^{(2+1)}$, as a result, factor $(y y_0)^{-1/2} e^{3t/8}$ was lost in [43].

More details on kernels $G^{(2+1)}$ and $G^{(3)}$, including derivation, are presented in the next Section.

3.3.2.1 Martingale Properties of Process *F*

The log-normal SABR ($\beta = 1$) is of a special interest, because it represents the borderline case between $\beta < 1$, when the SABR process F_t is a global martingale and $\beta > 1$, when F_t is a strict local martingale [53, 57, 71] (see also Chap. 4). For $\beta = 1$ process F_t proves to be a global martingale at $\rho \le 0$ and a strict local one at $\rho > 0$. We show this by computing directly the dimensionless first moment

$$
m_1(t) = \mathbb{E}\left[\frac{F_t}{F_0}\right] = \mathbb{E}\left[e^{Q_t - Q_0}\right]
$$

Despite already in possession of the invariant PDF $p_{\text{inv}}(q, V)$ (3.50), we have found more lucid to start with the driving SDE (3.1), writing it in the integral form

$$Q_t - Q_0 = \rho \int_0^t V_s \, \mathrm{d}W_s + \bar{\rho} \int_0^t V_s \, \mathrm{d}Z_s - \frac{1}{2} \int_0^t V_s^2 \, \mathrm{d}s \,,$$

where we used decomposition

$$W_t^1 = \rho W_t + \bar{\rho} Z_t, \qquad \mathbb{E}[\mathrm{d}W_t \, \mathrm{d}Z_t] = 0 \,.$$

The expectation of $\mathrm{e}^{Q_t - Q_0}$ is computed following the law of iterated expectations with probabilities of Brownian motions W and Z,

$$m_1(t) = \mathbb{E}_{\{W\}} \left[\exp\left\{ \rho \int_0^t V_s \, \mathrm{d}W_s - \frac{1}{2} \int_0^t V_s^2 \, \mathrm{d}s \right\} \right. \\ \left. \mathbb{E}_{\{Z\}} \left[\exp\left\{ \bar{\rho} \int_0^t V_s \, \mathrm{d}Z_s \right\} \middle| \, W_s, \; s < t \right] \right] .$$

Regarding the inner conditional expectation, we notice that first, processes V_s and Z_s are independent, and second, process V_s, conditioned on a given trajectory of W_s, may be treated as a deterministic function. Hence the inner expected value exists and is obtained by the Girsanov transform

$$\mathbb{E}_{\{Z\}} \left[\exp\left\{ \bar{\rho} \int_0^t V_s \, \mathrm{d}Z_s \right\} \middle| \, W_s, \; s < t \right] = \exp\left\{ \frac{\bar{\rho}^2}{2} \int_0^t V_s^2 \, \mathrm{d}s \right\} ,$$

resulting in

$$m_1(t) = \mathbb{E}_{\{W\}} \left[\exp\left\{ \rho \int_0^t V_s \, \mathrm{d}W_s - \frac{\rho^2}{2} \int_0^t V_s^2 \, \mathrm{d}s \right\} \right] . \tag{3.52}$$

First of all, notice that $m_1(t) = 1$ if $\rho = 0$ and proceed further with $\rho \neq 0$. The exponential in (3.52),

$$\mathscr{E}(\rho M)_t = \exp\left(\rho M_t - \tfrac{\rho^2}{2} (\mathrm{Var}\, M)_t \right) ,$$

is called the Doleans-Dade exponential of a local martingale $M_t = \int_0^t V_s \, \mathrm{d}W_s$ [52]. By Ito, $\mathscr{E}(\rho M)_t$ itself is a local martingale and a positive one (hence a supermartingale), meaning that

$$m_1(t) = \mathbb{E}_{\{W_2\}}[\mathscr{E}(\rho M)_t] \leq 1 \,.$$

Novikov's condition is not met in (3.52) [53, 60], and more refined methods are required to determine whether $\mathscr{E}(\rho M)_t$ is a global ($m_1 = 1$), or a strict local ($m_1 < 1$) martingale. It is interesting that the first moment $m_1(t)$ is completely determined by the dynamics of the volatility process V_t (and correlation ρ), while the underlying F_t is involved only in trivial scaling $(1/F_0)$. It was established in [5, 57, 60] that the martingale defect $1 - m_1$ can be expressed as the explosion probability of an auxiliary volatility process, with proper qualitative conclusions.

Here we proceed with direct calculation. According to driving equations (3.2, 3.3), the exponent in (3.52) can be rewritten as

$$\rho(V_t - V_0) - \frac{\rho^2}{2} I_t\,,$$

and the sought expectation can be found with the help of the join PDF $p(t;\, V, \tau \mid V_0)$ (3.6)

$$m_1(t) = \int e^{\rho(V_t - V_0) - \rho^2\tau/2}\, p(t;\, V, \tau \mid V_0)\, dV\, d\tau$$

(for brevity we will write merely p below). Now take derivative with respect to time and make use of the forward Kolmogorov equation, satisfied by p due to SDE-s (3.2, 3.3),

$$\partial_t p = -V^2 \partial_\tau p + \frac{1}{2}\partial_V^2 (V^2 p)\,, \tag{3.53}$$

so that

$$\partial_t m_1(t) = \int e^{\rho(V - V_0) - \rho^2\tau/2} \left(-V^2 \partial_\tau p + \frac{1}{2}\partial_V^2 (V^2 p)\right) dV\, d\tau\,.$$

Next, integrate term with $\partial_\tau p$ by parts and rearrange as follows,

$$\partial_t m_1(t) = \frac{1}{2} e^{-\rho V_0} \int e^{-\rho^2\tau/2} \left\{-\rho^2 e^{\rho V} V^2 p + e^{\rho V} \partial_V^2 (V^2 p)\right\} dV\, d\tau\,. \tag{3.54}$$

No end terms arise either at $\tau = 0$, or $\tau = \infty$ due to decaying exponentials $e^{-\rho^2\tau/2}$ at $\tau \to \infty$ and $e^{-(V^2 + V_0^2)/2\tau}$ in p at $\tau \to 0$. Noticing then, that $\rho^2 e^{\rho V} = \partial_V^2 e^{\rho V}$, rewrite expression in braces above as

$$\begin{aligned}\{\cdot\} &= \partial_V \left[-\partial_V (e^{\rho V}) V^2 p + e^{\rho V} \partial_V (V^2 p)\right] \\ &= \partial_V \left[e^{2\rho V} \partial_V (e^{-\rho V} V^2 p)\right]\end{aligned}$$

and integrate in (3.54) over V,

$$\partial_t m_1(t) = \frac{1}{2} e^{-\rho V_0} \int_0^\infty e^{-\rho^2\tau/2} \left[e^{2\rho V} \partial_V (e^{-\rho V} V^2 p)\right]_{V=0}^{V=\infty} d\tau\,.$$

At the lower limit $V = 0$, expression in brackets turns into zero owing to the properties of function $\theta(\frac{V V_0}{\tau}, t)$, which enters as a factor into PDF p. Namely, it can be shown (omitted) that $\theta(r, t)$ turns into zero faster, than any degree of r at r tending to zero. This leaves only the upper limit $V = \infty$.

At this point we plug in explicit expressions for PDF p (3.6) with θ (3.7), thus coming up with

$$\partial_t m_1(t) = \frac{e^{-\rho V_0 - t/8}}{2i\,(2\pi t)^{3/2}} \int_0^\infty e^{-\rho^2\tau/2}\,\frac{d\tau}{\tau}$$
$$\lim_{V\uparrow\infty}\left[e^{2\rho V}\partial_V\left(e^{-\rho V}(V V_0)^{1/2}\int_{C_u} e^{-u^2/2t - A^2/2\tau}\,u\,du\right)\right], \tag{3.55}$$
$$A^2 = V^2 + V_0^2 - 2V V_0\cosh u\,. \tag{3.56}$$

(factor A^2 is positive for all u on contour C_u, as discussed earlier in Sect. 3.2).

We may not take the limit of $V \to \infty$ first, because it would give zero (due to $e^{-V^2/2\tau}$), multiplied by the diverging integral over τ. We may, though, start with integration over τ. Gathering all factors, depending on τ, and making substitution $\tau = (A/|\rho|)e^w$, we get

$$J_{(\tau)} = \int_0^{+\infty} e^{-\rho^2\tau - A^2/2\tau}\,\frac{d\tau}{\tau} = \int_{-\infty}^{\infty} e^{-|\rho|A\,\cosh w}\,dw = 2K_0(|\rho|A)\,,$$

where $K_0(z)$ is the Macdonald function of the zeroth order [74], with known asymptotics at large arguments (large V means large A). In fact, the asymptotics of integral $J_{(\tau)}$ can be found directly by the Laplace's method without appealing to function K_0. In any case, at large A

$$J_{(\tau)} = \sqrt{\frac{2\pi}{|\rho|A}}\,e^{-|\rho|A}\,(1 + o(A^{-1/2}))\,,$$

and further expansion of A (3.56) as

$$A = V\left[1 - 2V_0/V\cosh u + (V_0/V)^2\right]^{1/2} = V - V_0\cosh u + o(V^{-1/2}),$$
$$A^{-1/2} = V^{-1/2}\,(1 + o(V^{-1/2}))$$

results in

$$J_{(\tau)} = \sqrt{\frac{2\pi}{|\rho|\,V}}\,e^{-|\rho|\,V + |\rho|\,V_0\cosh u}\,(1 + o(V^{-1/2}))\,.$$

Terms $o(V^{-1/2})$ do not affect the limit in V, and we drop them. Now substituting $J_{(\tau)}$ into (3.55) we obtain

$$\partial_t m_1(t) = \sqrt{\frac{\pi V_0}{2|\rho|}}\, \mathrm{e}^{-\rho V_0 - t/8} \lim_{V\uparrow\infty} \left[\mathrm{e}^{2\rho V}\partial_V \left(\mathrm{e}^{-\rho V - |\rho| V}\right)\right]$$
$$\times \frac{1}{\mathrm{i}\,(2\pi t)^{3/2}} \int_{C_u} \mathrm{e}^{|\rho|\, V_0 \cosh u - u^2/2t}\, u\, \mathrm{d}u\,.$$

Remarkably, the contribution from the limit in V and from the integral over u is factorized, moreover, the integral generates the same function $\theta(r)$ (3.7) with $r = |\rho|\, V_0$.

Turning to the limit, we notice that if $\rho < 0$, then $\lim_{V\uparrow\infty} \left[\mathrm{e}^{-2|\rho| V}\partial_V(1 + o(V^{-1/2}))\right] = 0$, accordingly $\partial_t m_1(t) = 0$, and $m_1(t) = m_1(0) = 1$. Process F_t is the global martingale at $\rho \le 0$.

If $\rho > 0$, then

$$\lim_{V\uparrow\infty} \left[\mathrm{e}^{2\rho V}\partial_V \left(\mathrm{e}^{-2\rho V}\right)\right] = -2\rho\,,$$

and we come up with the new compact expression for the $m_1(t)$ rate of change

$$\partial_t m_1(t) = -\sqrt{2\pi\rho V_0}\, \mathrm{e}^{-\rho V_0 - t/8}\, \theta(\rho V_0) < 0\,. \qquad (3.57)$$

Note that function $\theta(r, t)$ enters as a factor into the joint PDF (3.5) and is always positive, though it is difficult to draw from its integral representation (3.7).

Integration over time gives $m_1(t) < 1$ for $\rho > 0$. The ultimate value $m_1(\infty)$ takes especially simple form. Using $\theta(r, t)$ (3.7) and integrating over time from 0 to ∞, we have (denote for brevity $r = \rho V_0$)

$$m_1(\infty) - 1 = -\frac{\sqrt{2\pi r}\,\mathrm{e}^{-r}}{2\pi \mathrm{i}} \int_{C_u} \mathrm{e}^{r\cosh u}\, \mathrm{d}u \left(\frac{u}{\sqrt{2\pi}} \int_0^\infty \mathrm{e}^{-t/8 - u^2/2t}\, t^{-3/2}\, \mathrm{d}t\right).$$

The inner integral over t is of the type, we have already met, (3.21), and is equal to $\mathrm{e}^{-u/2}$, hence

$$m_1(\infty) - 1 = -\sqrt{2\pi r}\,\mathrm{e}^{-r} \left(\frac{1}{2\pi \mathrm{i}} \int_{C_u} \mathrm{e}^{r\cosh u - u/2}\, \mathrm{d}u\right).$$

Contour integral is exactly the representation of the Bessel function $I_{1/2}(r)$ (look at (2.57)), and the latter is expressed through elementary functions [74], $I_{1/2}(r) = \sqrt{\frac{2}{\pi r}} \sinh r$. As a result (obtained also in [57, 60]),

$$m_1(\infty) = 1 - 2\mathrm{e}^{-r}\, \sinh r = \mathrm{e}^{-2r} = \mathrm{e}^{-2\rho V_0}\,.$$

Using another approach, related to explosion of the volatility process V_t, Lewis [57] obtained expression for moment $m_1(t)$ at finite times t, with time derivative $\partial_t m_1(t)$, that is visually absolutely different from ours (3.57). Since both are claimed the new results, it is instructive to check that they are identical. To do this, we transform function θ (3.7) as follows. Recall (3.7),

$$\theta(r,t) = \frac{1}{\mathrm{i}\,(2\pi t)^{3/2}} \int_{C_u} \mathrm{e}^{r\cosh u - u^2/2t} u\,\mathrm{d}u\,. \tag{3.58}$$

Start with factor $u\,\mathrm{e}^{-u^2/2t}$ and present Gaussian exponential in the form of Fourier integral (which works also for complex u on contour C_u)

$$\begin{aligned} u\,\mathrm{e}^{-u^2/2t} &= -t\,\partial_u \mathrm{e}^{-u^2/2t} = -t\sqrt{\frac{t}{2\pi}}\,2\partial_u \int_0^\infty \mathrm{e}^{-s^2t/2}\cos(su)\,\mathrm{d}s \\ &= 2\frac{t^{3/2}}{\sqrt{2\pi}} \int_0^\infty \mathrm{e}^{-s^2t/2}\sin(su)\,s\,\mathrm{d}s\,. \end{aligned}$$

Substitute the last expression into θ (3.58) and change the order of integration,

$$\theta(r,t) = \frac{2}{\mathrm{i}\,(2\pi)^2} \int_0^\infty \mathrm{e}^{-s^2t/2}\,s\,\mathrm{d}s \int_{C_u} \mathrm{e}^{r\cosh u}\sin(su)\,\mathrm{d}u\,.$$

The contour integral is computed as

$$\begin{aligned} \int_{C_u} (\cdot)\,\mathrm{d}u &= \frac{1}{2\mathrm{i}} \int_{C_u} \mathrm{e}^{r\cosh u}(\mathrm{e}^{\mathrm{i}su} - \mathrm{e}^{-\mathrm{i}su})\,\mathrm{d}u \\ &= \pi[I_{-\mathrm{i}s}(r) - I_{\mathrm{i}s}(r)] = 2\mathrm{i}\,\sinh(s\pi)\,K_{\mathrm{i}s}(r) \end{aligned}$$

due to the integral representation (2.57) of Bessel functions and according to definition of Macdonald function $K_\nu(r)$ [1]. This results in the another expression for function θ

$$\theta(r,t) = \frac{1}{\pi^2} \int_0^\infty \sinh(s\pi)\,K_{\mathrm{i}s}(r)\,\mathrm{e}^{-s^2t/2}\,s\,\mathrm{d}s\,.$$

Plugging this into (3.57) for $\partial_t m_1(t)$ with $\rho V_0 = r$ and integrating over time, we obtain expression

$$m_1(\infty) - m_1(t) = -\left(\frac{2}{\pi}\right)^{3/2} \sqrt{r}\,\mathrm{e}^{-r-t/8} \int_0^\infty \sinh(s\pi)\,K_{\mathrm{i}s}(r)\mathrm{e}^{-s^2t/2} \frac{s\,\mathrm{d}s}{s^2+\frac{1}{4}}\,,$$

equivalent to one in [57], Chap. 8.4. Plugging this into (3.57) for $\partial_t m_1(t)$ with $\rho V_0 = r$ and integrating over time, we obtain expression from [57], Chap. 8.4

$$m_1(\infty) - m_1(t) = -\left(\frac{2}{\pi}\right)^{3/2} \sqrt{r}\,\mathrm{e}^{-r-t/8} \int_0^\infty \sinh(s\pi)\,K_{\mathrm{i}s}(r)\mathrm{e}^{-s^2t/2} \frac{s\,\mathrm{d}s}{s^2+\frac{1}{4}}\,.$$

3.4 McKean and Related Heat Kernels

In this Section we provide a new derivation of the heat kernels, associated with the normal and log-normal SABR models, as solutions of the corresponding heat (Kolmogorov) equations. We combined derivations in the separate section, as they follow one from another.

3.4.1 *Basics of Hyperbolic Geometry*

We start with some basic facts of the geometry of the hyperbolic space $\mathbf{H}^n$. There are $n-1$ 'horizontal' coordinates x_α and one 'vertical' y. For brevity denote $|x|$ the Euclidean length of vector x_α in $\mathbf{R}^{n-1}$ and $\mathrm{d}x = \Pi \mathrm{d}x_\alpha$. Geometric elements in $\mathbf{H}^n$ are

$$\delta l^2 = \frac{1}{y^2}\left(|\mathrm{d}x|^2 + \mathrm{d}y^2\right), \quad g_{ij} = y^{-2} I_{ij}, \quad g = y^{-2n},$$
$$\mathrm{d}V_{\mathrm{inv}} = y^{-n} \mathrm{d}x \mathrm{d}y, \qquad \delta_{\mathrm{inv}}(z, z') = y^n \delta(z - z')\,.$$

Laplace-Beltrami operator

$$\Delta = g^{-1/2} \partial_i g^{1/2} g^{ij} \partial_j = y^2 \left(\textstyle\sum_\alpha \partial^2_{x_\alpha} + \partial^2_y + (2-n) y^{-1} \partial_y\right). \tag{3.59}$$

Hyperbolic distance $s(z, z')$ is defined by

$$\mu = \cosh s = \frac{|x - x'|^2 + y^2 + y'^2}{2yy'}\,. \tag{3.60}$$

The 'radial' part of Laplacian is the form it takes when acting on a function depending only on distance s; it is convenient to use $\mu = \cosh s$. Like in any metric space,

$$\begin{aligned}\Delta f(\mu) &= \mathrm{div}\,\mathrm{grad}(f(\mu)) = \mathrm{div}\left(\partial_\mu f \;\; \mathrm{grad}\,\mu\right)\\ &= (\mathrm{grad}\,\mu)^2\, \partial^2_\mu f + (\Delta\mu)\, \partial_\mu f\,.\end{aligned}$$

Then follows exercise in differentiation, based on (3.60),

$$\begin{aligned}(\mathrm{grad}\,\mu)^2 &= g^{ij}(\partial_i \mu)(\partial_j \mu) = y^2\left(\textstyle\sum_\alpha (\partial_{x_\alpha}\mu)^2 + (\partial_y \mu)^2\right) = \mu^2 - 1\,,\\ \Delta^{(n)}\mu &= y^2 \textstyle\sum_\alpha (\partial^2_{x_\alpha}\mu) + y^n \partial_y (y^{2-n} \partial_y \mu) = n\mu\,,\end{aligned}$$

yielding the 'radial' part of Laplacian

$$\Delta^{(n)} f(\mu) = (\mu^2 - 1)\, \partial^2_\mu f \; + n\mu\, \partial_\mu f\,. \tag{3.61}$$

On account of $\mu = \cosh s$, $\partial_\mu = (\sinh s)^{-1}\partial_s$, it can also be expressed directly in terms of s

$$\Delta^{(n)} f(s) = \frac{1}{(\sinh s)^{n-1}} \frac{\partial}{\partial s} \left[(\sinh s)^{n-1} \frac{\partial f}{\partial s} \right] . \tag{3.62}$$

The last of preliminary formulas is the commutation relation between 'radial' Laplacian and differentiation operator ∂_μ. Differentiating Laplacian of a function $f(\mu)$, we get

$$\partial_\mu \, \Delta^{(n)} f(\mu) = \partial_\mu \, \left((\mu^2 - 1) \, \partial_\mu^2 + n\mu \, \partial_\mu \right) f(\mu) \tag{3.63}$$

$$= \left((\mu^2 - 1) \, \partial_\mu^2 + (n+2)\mu\partial_\mu + n \right) \partial_\mu f(\mu) \tag{3.64}$$

$$= \left(\Delta^{(n+2)}(\mu) + n \right) \partial_\mu f(\mu) . \tag{3.65}$$

This allows to generate consecutively (with step two) higher heat kernels.

3.4.2 *Invariant Heat Kernels* $G^{(n)}$

Invariant heat kernel $G^{(n)}$ is the fundamental solution of the heat equation in $\mathbf{H}^n$

$$\partial_t G^{(n)} = \frac{1}{2} \Delta^{(n)} \, G^{(n)} \tag{3.66}$$

with initial distribution

$$G^{(n)}_{|\, t=0} = \delta^{(n)}_{\text{inv}}(z, z_0) .$$

It proves by construction that so defined heat kernels $G^{(n)}$ depend only on time t and the hyperbolic distance $s(z, z_0)$ (or on $\mu = \cosh s$) in any dimension n. This property admitted, kernels $G^{(n)}$ solve the heat equation (3.66) with 'radial' Laplacian (3.61). This allows to differentiate equation (3.66) with respect to μ and to make use of the found commutation relation (3.65),

$$\partial_t \, \partial_\mu G^{(n)} = \frac{1}{2} \partial_\mu \, \Delta^{(n)} \, G^{(n)} = \frac{1}{2} \Delta^{(n+2)} \, \partial_\mu G^{(n)} + \frac{n}{2} \, \partial_\mu G^{(n)} .$$

Absorbing the last term on the right side into the time derivative leads to the heat equation in $\mathbf{H}^{n+2}$

$$\partial_t \, [\mathrm{e}^{-\frac{nt}{2}} \, \partial_\mu G^{(n)}] = \frac{1}{2} \Delta^{(n+2)} \, [\mathrm{e}^{-\frac{nt}{2}} \, \partial_\mu G^{(n)}]$$

Its solution $e^{-\frac{nt}{2}}\partial_\mu G^{(n)}$ depends only on t and μ and is proportional to $G^{(n+2)}$ with constant coefficient to be found from normalization conditions. Exact relation looks like [63]

$$G^{(n+2)} = -\frac{1}{2\pi}e^{-\frac{nt}{2}}\partial_\mu G^{(n)} = -\frac{1}{2\pi}\frac{e^{-\frac{nt}{2}}}{\sinh s}\partial_s G^{(n)} \tag{3.67}$$

and allows obtaining all 'odd' heat kernels from $G^{(1)}$ and all 'even' ones from $G^{(2)}$.[4]

This tells little, though, how to find the important McKean's heat kernel $G^{(2)}$. We have found a new, quite elementary derivation; no special functions, no contour integrals, only some algebra and differentiation. It follows scheme $G^{(1)} \to G^{(3)} \to G^{(2+1)} \to G^{(2)}$ where $G^{(1)}$ (see below) is Gaussian distribution, $G^{(3)}$ is obtained from $G^{(1)}$ by simple differentiation, as discussed above; and new steps, kernel $G^{(2+1)}$, that solves a heat equation arising in connection with log-normal SABR model, and modifies $G^{(3)}$ multiplying it by function $(yy_0)^{-1/2}\exp(\frac{3}{8}t)$; finally, $G^{(2)}$ follows as a marginal distribution from $G^{(2+1)}$ by integrating out one of two 'horizontal' coordinates.

We start now with **Dimension one**. Element of length is $\delta l^2 = ds^2 = y^{-2}dy^2$, distance $s = \log(y/y_0)$ (for $n = 1$ it is better to take distance with sign). Laplacian (3.62) looks like $\Delta^{(1)} = \partial_s^2$. The heat equation becomes the standard parabolic one

$$\partial_t G^{(1)} = \tfrac{1}{2}\partial_s^2 G^{(1)},$$

its fundamental solution

$$G^{(1)}(t,s) = \frac{1}{\sqrt{2\pi t}}e^{-\frac{s^2}{2t}}$$

is, in fact, the Gaussian distribution for a log-normal process with initial condition

$$G^{(1)}_{|t=0} = \delta(s) = \delta\left(\log(y/y_0)\right) = y_0\delta(y-y_0) = \delta^{(1)}_{\text{inv}}(y,y_0),$$

meaning that it is the invariant heat kernel in $\mathbf{H}^1$.

Dimension three. There are two horizontal coordinates x_1, x_2 and one vertical y. Distance u, and Laplacian $\Delta^{(3)}$ are

$$\cosh u = \frac{x_1^2 + x_2^2 + y^2 + y_0^2}{2yy_0}, \tag{3.68}$$

$$\Delta^{(3)} = y^2(\partial_{x_1}^2 + \partial_{x_2}^2 + \partial_y^2 - y^{-1}\partial_y). \tag{3.69}$$

We have chosen notation u for the geodesic distance in $\mathbf{H}^3$, reserving now notation s for dimension two. The heat kernel $G^{(3)}$ is obtained from (3.67) with $n = 1$,

[4]Consideration of spaces $\mathbf{H}^n$ starts usually from $n = 2$. We observed, however, that space $\mathbf{H}^1$ is also admissible.

$$G^{(3)}(t,u) = -\frac{1}{2\pi}\frac{e^{-\frac{t}{2}}}{\sinh u}\partial_u G^{(1)}(t,u) = \frac{e^{-\frac{t}{2}}}{(2\pi t)^{3/2}}\frac{u}{\sinh u}e^{-\frac{u^2}{2t}}. \tag{3.70}$$

It can be verified directly, that $G^{(3)}(t,u)$ is properly normalized

$$G^{(3)}_{|t=0} = \delta^{(3)}_{\text{inv}}(z,z_0) = y_0^3\,\delta(x_1)\delta(x_2)\delta(y-y_0).$$

3.4.3 Noncanonical Kernel $G^{(2+1)}$

Function $G^{(2+1)}$ arises in Sect. 3.3.2 in connection with the log-normal SABR model and solves diffusion equation (3.46)

$$\partial_t G^{(2+1)} = \tfrac{1}{2}y^2(\partial^2_{x_1} + \partial^2_{x_2} + \partial^2_y)G^{(2+1)} \tag{3.71}$$

with the postulated initial condition

$$G^{(2+1)}_{|t=0} = y_0^2\delta(x_1)\delta(x_2)\delta(y-y_0) = y_0^{-1}\delta^{(3)}_{\text{inv}}(z,z_0). \tag{3.72}$$

Factor $\delta(x_2)$ in (3.72) guarantees that Fourier transform of the initial condition of $G^{(2+1)}$ coincide with the initial condition (3.44) of its Fourier transform $\tilde{p}$ (3.45).

Compared with the true Laplacian $\Delta^{(3)}$ (3.69), diffusion operator in (3.71) (call it $\Delta^{(2+1)}$) lacks the term with the first derivative. The way to solve the last equation is to transform quasi-Laplacian $\Delta^{(2+1)}$ into the true one $\Delta^{(3)}$. Of course it is possible to invoke the general methodics [43, 67], but the task is simple enough to be treated in a simple way. It is almost obvious that substitution $G^{(2+1)} \sim y^{-1/2}\tilde{G}$ generates the sought differential structure $\partial^2_y - y^{-1}\partial_y$. Indeed, proceeding with substitution

$$G^{(2+1)} = (yy_0)^{-1/2}\tilde{G}, \tag{3.73}$$

we transform Eq. (3.71) into the following one for $\tilde{G}$ (factor $y_0^{-1/2}$ is used for convenience)

$$\begin{aligned}\partial_t\tilde{G} &= \tfrac{1}{2}y^2(\partial^2_{x_1} + \partial^2_{x_2} + \partial^2_y - y^{-1}\partial_y + \tfrac{3}{4}y^{-2})\tilde{G}\\ &= \tfrac{1}{2}\Delta^{(3)}\tilde{G} + \tfrac{3}{8}\tilde{G}.\end{aligned}$$

Term $\frac{3}{8}\tilde{G}$ is readily absorbed into the time derivative thus leading to

$$\partial_t(\tilde{G}e^{-\frac{3}{8}t}) = \tfrac{1}{2}\Delta^{(3)}(\tilde{G}e^{-\frac{3}{8}t}),$$

which is nothing but the heat equation in $\mathbf{H}^3$. Next, the initial distribution of the solution $\tilde{G}e^{-\frac{3}{8}t}$ is related to that of $G^{(2+1)}$ due to (3.73) and (3.72)

$$(\tilde{G}e^{-\frac{3}{8}t})_{|t=0} = \tilde{G}_{|t=0} = (yy_0)^{\frac{1}{2}}\, G^{(2+1)}_{|t=0} = \delta^{(3)}_{\text{inv}}(z, z_0)\,.$$

We conclude that function $\tilde{G}e^{-\frac{3}{8}t}$ coincides with the invariant heat kernel $G^{(3)}$ (3.70). Accordingly, the sought quasi-kernel $G^{(2+1)}$ is ultimately found in the form

$$G^{(2+1)}(t, x_1, x_2, y) = \frac{e^{\frac{3}{8}t}}{\sqrt{yy_0}} G^{(3)}(t, u) = \frac{e^{-\frac{1}{8}t}}{(2\pi t)^{3/2}\sqrt{yy_0}} \frac{u}{\sinh u} e^{-\frac{u^2}{2t}} \tag{3.74}$$

with distance u defined by (3.68).

Being itself of importance when dealing with the log-normal SABR, function $G^{(2+1)}$ serves as the last intermediate result in seeking the McKean heat kernel $G^{(2)}$.

3.4.4 *McKean Heat Kernel* $G^{(2)}$

In **Dimension two** there are one horizontal x and one vertical coordinate y. Geometric elements are already defined by (3.33, 3.34, 3.37), we recall only that distance in $\mathbf{H}^2$ looks like

$$\cosh s(z, z_0) = \frac{(x - x_0)^2 + y^2 + y_0^2}{2yy_0}. \tag{3.75}$$

The invariant heat kernel $G^{(2)}(t, s)$ solves the heat equation in $\mathbf{H}^2$

$$\partial_t G^{(2)} = \frac{1}{2} y^2 (\partial_x^2 + \partial_y^2) G^{(2)} \tag{3.76}$$

with initial distribution

$$G^{(2)}_{|t=0} = y_0^2 \delta(x - x_0)\delta(y - y_0)\,. \tag{3.77}$$

Now look back at the equation and the initial condition for G^{2+1} (3.71 and 3.72), and integrate both over all real x_2 (equally take twice the value of integral over positive x_2).

It is obvious that function

$$\begin{aligned}\tilde{G}^{(2)}(t, x - x_0, y; y_0) &= 2\int_0^\infty G^{(2+1)}(t, x - x_0, x_2, y)\, dx_2 \\ &= \frac{2\, e^{-t/8}}{(2\pi t)^{3/2}\sqrt{yy_0}} \int_0^\infty \frac{u}{\sinh u} e^{-\frac{u^2}{2t}}\, dx_2 \end{aligned} \tag{3.78}$$

solves the $2D$ heat equation (3.76) with proper initial condition (3.77) and must therefore coincide with McKean kernel $G^{(2)}(t, s)$ (3.15). To show this, simply change variable of integration in the last integral from x_2 to u. Comparing expressions (3.68) and (3.75) for 3D distance u (with $x_1 = x - x_0$) and for 2D distance s, we get

$$x_2 = \sqrt{2yy_0}\sqrt{\cosh u - \cosh s}, \qquad \mathrm{d}x_2 = \frac{1}{2}\sqrt{2yy_0}\frac{\sinh u \,\mathrm{d}u}{\sqrt{\cosh u - \cosh s}}.$$

Making this substitution in (3.78) immediately transforms $\tilde{G}^2$ into the known McKean's expression (3.15)

$$\tilde{G}^2 \equiv G^{(2)}(t,s) = \frac{\sqrt{2}\,\mathrm{e}^{-t/8}}{(2\pi t)^{3/2}} \int_s^\infty \frac{u\,\mathrm{e}^{-\frac{u^2}{2t}}}{\sqrt{\cosh u - \cosh s}}\,\mathrm{d}u\,. \tag{3.79}$$

Note that all we have used in this derivation are few formulas from hyperbolic geometry (distance, Laplacian, invariant delta function).

3.4.5 Cumulative Distribution Function $G(t, s)$

With the invariant PDF for the Brownian diffusion on $\mathbf{H}^2$ given by $G^2(t, s)$, the associated cumulative distribution function (CDF) $G(t, s)$ is defined as

$$G(t,s) = \int G^{(2)}(t, s(x,y))\mathbf{1}_{s(x,y)>s}\,\frac{\mathrm{d}x\,\mathrm{d}y}{y^2}, \tag{3.80}$$

and describes the probability $P(s(x, y) > s)$ to find the observation point (x, y) at distances, greater than s. When integrating over x, we may put $x_0 = 0$ in (3.75) and then take only positive x-s (twice). Changing variables of integration $x \to s' = s(x, y)$ and $y \to w = \log(y/y_0)$, we rewrite CFD G as

$$G(t,s) = \sqrt{2}\int_s^\infty G^{(2)}(t,s')\,\sinh s'\mathrm{d}s' \int_{-s'}^{s'} \frac{\mathrm{e}^{-w/2}}{\sqrt{\cosh s' - \cosh w}}\,\mathrm{d}w\,.$$

Keeping only the even part of exponential $\mathrm{e}^{-w/2} \to \cosh(w/2$ and passing to variable $r = \sinh(w/2)$ transforms integral over w into an elementary one, resulting altogether in the following definition of CDF

$$G(t,s) = 2\pi \int_s^\infty G^{(2)}(t,s')\,\sinh s'\,\mathrm{d}s'\,, \tag{3.81}$$

which looks quite obvious with somewhat more advanced knowledge of $\mathbf{H}^2$ geometry. Namely, it is possible to introduce in $\mathbf{H}^2$ polar coordinates, distance s and angle φ, with element of volume $\mathrm{d}V_{\mathrm{inv}} = \sinh s\,\mathrm{d}s\,\mathrm{d}\varphi$, thus making transition from (3.80) to (3.81) trivial.

Now plugging $G^{(2)}(t, s)$ (3.79) into (3.81), changing the order of integration, and integrating over s', we come up with two equivalent expressions

$$G(t, s) = \frac{2\,\mathrm{e}^{-t/8}}{t\sqrt{\pi t}} \int_s^\infty \mathrm{e}^{-\frac{u^2}{2t}} u \sqrt{\cosh u - \cosh s}\, du\,, \tag{3.82}$$

$$= \frac{\mathrm{e}^{-t/8}}{\sqrt{\pi t}} \int_s^\infty \frac{\mathrm{e}^{-\frac{u^2}{2t}} \sinh u}{\sqrt{\cosh u - \cosh s}}\, du\,. \tag{3.83}$$

Consider asymptotics of CDF $G(t, s)$ for large s to be used in the option valuation. Making substitution

$$u = (s^2 + w^2)^{1/2} \simeq s + \frac{w^2}{2s}$$

and keeping the leading order in w/s in each factor, we find after integration over w

$$G(t, s) \simeq \exp\left(-\frac{t}{8} - \frac{s^2}{2t}\right) \sqrt{\frac{\sinh s}{s}}\,. \tag{3.84}$$

A better approximation for $G(t, s)$, that works in a wide range of arguments, is presented in Sect. 3.7.3 (3.119) [9, 10].

CDF $G(t, s)$ appears in most expressions for option values.

3.5 Option Value for the Free Normal SABR

We start with the forward Kolmogorov equation for PDF $p(F, V)$ of F, V

$$\partial_t p(F, V) = \frac{1}{2}\left(\partial_F^2 + 2\rho\partial_F\partial_V + \partial_V^2\right) V^2 p(F, V) \tag{3.85}$$

and integrate it with payoff $(F - K)^+$ over the full space (F, V). In the right hand side, only the the first term (with ∂_F^2) contributes, we integrate it twice by parts with respect to F to get the time derivative of the option in the form

$$\partial_t C(K, V) = \frac{1}{2} \int_0^\infty V^2 p(K, V)\, dV\,.$$

PDF $p(F, V)$ relates to the invariant PDF $p_{\text{inv}} = G^{(2)}$, as

$$p_{\text{inv}} = \frac{p(F, V)}{\sqrt{g(F, V)}} = \bar{\rho} V^2 p(F, V)\,,$$

resulting in

$$\partial_t C(t,K) = \frac{1}{2\bar{\rho}} \int_0^\infty G^{(2)}(t, s(K,V))\mathrm{d}V\,. \tag{3.86}$$

Distance s is defined by (3.37) with 'orthogonal' coordinates x, y, then expressed through original ones F, V in (3.38), and is taken at $F = K$.

3.5.1 Expression Using CDF $G(t, s)$

The last integral can be transformed using integration by parts and passing to s as a variable of integration. First, express $\cosh s$ (3.37) in terms of $F = K$ and V,

$$\cosh s = \frac{1}{2VV_0}\left[\frac{1}{\bar{\rho}^2}(K - \rho V - F_0 + \rho V_0)^2 + V^2 + V_0^2\right].$$

Introduce scaled strike k ,

$$k = \frac{K - F_0}{V_0} + \rho \tag{3.87}$$

and expand $\cosh s$ in powers of V,

$$\cosh s = \frac{1}{\bar{\rho}^2}\left(\frac{(k^2 + \bar{\rho}^2)V_0}{2V} + \frac{V}{2V_0} - \rho k\right). \tag{3.88}$$

Next, parametrize V as

$$V = V_0\sqrt{k^2 + \bar{\rho}^2}e^w\,,$$

thus presenting $\cosh s$ as a function of w and vice versa (note though that s is positive while w may take both signs)

$$\cosh s = \frac{1}{\bar{\rho}^2}\left(\sqrt{k^2 + \bar{\rho}^2}\cosh w - \rho k\right). \tag{3.89}$$

Adopting w as a new variable of integration we get

$$\partial_t C(t,K) = \frac{V_0\sqrt{k^2 + \bar{\rho}^2}}{2\bar{\rho}} \int_{-\infty}^\infty \mathrm{d}w\, \mathrm{e}^w G^{(2)}(t, s(w))\,.$$

Taking into account that s is an even function of w we keep only the even part in the exponential $e^w \to \cosh w$, then present the above integral as $\int_{-\infty}^\infty \mathrm{d}w(...) = 2\int_0^\infty \mathrm{d}w(...)$ and integrate by parts

$$2\int_0^\infty \cosh w\, G^{(2)}(t, s(w))\,\mathrm{d}w = -2\int_{s_0}^\infty \sinh w(s)\, \partial_s G^{(2)}(t,s)\,\mathrm{d}s\,,$$

where in the last passage we have switched to s as variable of integration. The lower limit s_0 is obtained from expression (3.89) for $\cosh s$ at $w = 0$

$$\cosh s_0 = \frac{1}{\bar{\rho}^2}\left(\sqrt{k^2+\bar{\rho}^2} - \rho k\right) \tag{3.90}$$

Then using relation (3.41) between $\partial_s G^{(2)}$ and $\partial_t G(t,s)$ we collect pieces to get

$$\partial_t C(t,K) = \frac{V_0\sqrt{k^2+\bar{\rho}^2}}{\pi\bar{\rho}} \int_{s_0}^{\infty} \frac{\sinh w(s)}{\sinh s}\, \partial_t G(t,s)\, \mathrm{d}s$$

and integrate over time to obtain the option time value

$$\begin{aligned}\mathcal{O}(t,K) &= C(t,K) - (F_0 - K)^+ \\ &= \frac{V_0\sqrt{k^2+\bar{\rho}^2}}{\pi\bar{\rho}} \int_{s_0}^{\infty} \frac{\sinh w(s)}{\sinh s}\, G(t,s)\, \mathrm{d}s\end{aligned}$$

(we have taken into account that $G(0,s) = 0$ because at $t = 0$ probability density $G^{(2)}(0,s)$ is delta-concentrated at $s = 0$ implying that cumulative probability of being outside any distance $s \geq s_0 > 0$ is zero).

Finally, express $\sinh w$ through s with the help of (3.89)

$$\begin{aligned}\sinh^2 w &= \frac{(\bar{\rho}^2\cosh s + \rho k)^2 - k^2 - \bar{\rho}^2}{k^2+\bar{\rho}^2} \\ &= \frac{\bar{\rho}^2}{k^2+\bar{\rho}^2}\left[\sinh^2 s - (k - \rho\cosh s)^2\right]\end{aligned}$$

Summarizing, The option time value for the free normal SABR model with nonzero correlation and with rate F not restricted by sign is

$$\mathcal{O}(t,K) = \frac{V_0}{\pi}\int_{s_0}^{\infty} \frac{G(t,s)}{\sinh s}\sqrt{\sinh^2 s - (k - \rho\cosh s)^2}\, \mathrm{d}s \tag{3.91}$$

with cumulative probability $G(t,s)$, scaled strike k, and lower limit s_0 given by (3.82), (3.87), and (3.90).

3.5.1.1 Zero Correlation

When ρ turns into zero, $\bar{\rho} = 1$, $k = \frac{K-F_0}{V_0}$, $\cosh s_0 = \sqrt{k^2+1}$, $\sinh s_0 = |k|$, and the last expression simplifies to

$$\mathcal{O}_N(t, s_0) = \frac{V_0}{\pi} \int_{s_0}^{\infty} \frac{G(t,s)}{\sinh s} \sqrt{\sinh^2 s - \sinh^2 s_0}\, \mathrm{d}s\,, \qquad \sinh s_0 = \frac{|K - F_0|}{V_0}\,. \tag{3.92}$$

We have introduced this notation for the time value to indicate that it refers to the normal SABR and to underline its dependence on s_0 (also keeping in mind to use it in the derivation of the log-normal option value).

In addition, if only positive F are allowed, the invariant PDF (absorbing) is given by the difference of direct and reflected terms (3.16),

$$p_{\text{inv}} = G^{(2)}(t, s_d) - G^{(2)}(t, s_r),$$

accordingly, the time option value is obtained as the difference of two integrals

$$\mathcal{O}(t, K) = \mathcal{O}_N(t, s_-) - \mathcal{O}_N(t, s_+) \qquad \sinh s_\pm = \frac{|K \pm F_0|}{V_0}\,. \tag{3.93}$$

Notice that only the 'direct' term contributes to the intrinsic option value, while for the 'reflected' one, $(-F_0 - K)^+ = 0$.

The obtained expression is for the classic normal SABR with the absorbing boundary at zero. For reflecting boundaries the option time value will be

$$\mathcal{O}(t, K) = \mathcal{O}_N(t, s_-) + \mathcal{O}_N(t, s_+)\,.$$

3.5.2 Alternative Expression Using the Normal CDF

A somewhat different approach was employed by Henry-Labordère (HL) [43] and then by Korn and Tang (KT) [54]. They used the integral representation (3.15) for McKean kernel $G^{(2)}(t, s)$, rather than treating $G^{(2)}(t, s)$ as a given function. The time derivative (3.86) is then written as a two dimensional integral

$$\partial_t C(t, K) = \frac{\sqrt{2} V_0\, e^{-t/8}}{2\bar{\rho}(2\pi t)^{3/2}} \int_0^{\infty} \mathrm{d}v \int_{s(v)}^{\infty} \frac{u\, \mathrm{e}^{-\frac{u^2}{2t}}}{\sqrt{\cosh u - \cosh s(v)}}\, \mathrm{d}u\,, \tag{3.94}$$

where $v = \frac{V}{V_0}$ and $\cosh s(v)$ is defined in (3.88). HL tried to interchange the order of integration over V and u, however, claiming erroneously that the region of integration in the plane (v, u) was a rectangular half-strip (which it is not) led the author to the wrong answer. The mistake was pointed out by KT who also performed integration with respect to time, thus presenting the option value in the form of a two dimensional integral [54]. We take further step in this direction performing integration with respect to v and presenting the option value as one dimensional integral (different from the previous one (3.91)). Functions involved are the normal cumulative probability

$$\Phi(x) = \frac{1}{\sqrt{2\pi}} \int_{-\infty}^{x} e^{-\frac{\xi^2}{2}} d\xi$$

and the full elliptic integral of the second kind

$$E(m) = \int_0^{\pi/2} \sqrt{1 - m \sin^2 \varphi d\varphi}\,, \tag{3.95}$$

which is a standard special function, even though less common than $\Phi(x)$.

Turning to double integral (3.94) notice that the domain of integration in plane (v, u) is defined by condition

$$\cosh u \geq \cosh s(v)\,.$$

According to expression (3.88) for $\cosh s$, variable u has a global minimum $u = s_0$ with $\cosh s_0$ given by (3.90). The above inequality can be rewritten as

$$0 \geq \cosh s(v) - \cosh u = \frac{1}{2\bar{\rho}^2 v} \left[v^2 - 2(\bar{\rho}^2 \cosh u + \rho k)v + k^2 + \bar{\rho}^2 \right].$$

If integration with respect to v is performed first, variable v will be confined in interval

$$v_- \leq v \leq v_+\,,$$

where $v_\pm(u)$ are roots of the quadratic polynomial in brackets,

$$v_\pm(u) = \bar{\rho}^2 \cosh u + \rho k \pm \bar{\rho} \sqrt{\sinh^2 u - (k - \rho \cosh u)^2}\,. \tag{3.96}$$

As a result, integral (3.94) is transformed to

$$\partial_t C(t, K) = \frac{\sqrt{2} V_0\, e^{-t/8}}{2\bar{\rho}(2\pi t)^{3/2}} \int_{s_0}^{\infty} du\, u\, e^{-\frac{u^2}{2t}} \int_{v_-}^{v_+} \frac{dv}{\sqrt{\cosh u - \cosh s(v)}}\,.$$

We transform the inner integral into the standard elliptic integral (3.95). First, using roots $v_\pm$ we express

$$\cosh u - \cosh s(v) = \frac{(v - v_-)(v_+ - v)}{2\bar{\rho}^2 v}\,,$$

coming up with

$$J_{(v)} = \int_{v_-(u)}^{v_+(u)} \frac{dv}{\sqrt{\cosh u - \cosh s(v)}} = \sqrt{2}\bar{\rho} \int_{v_-(u)}^{v_+(u)} \frac{\sqrt{v}\, dv}{\sqrt{(v - v_-)(v_+ - v)}}\,. \tag{3.97}$$

Presence of three square roots clearly indicates that integral $J_{(v)}$ is related to elliptic functions. Indeed, by introducing a new variable of integration

$$w = \sqrt{\frac{v_+ - v}{v_+ - v_-}}$$

that changes from $w = 1$ $(v = v_-)$ to $w = 0$ $(v = v_+)$, we recast $J_{(v)}$ into

$$J_{(v)} = 2\bar{\rho}\sqrt{2v_+}\int_0^1 \sqrt{\frac{1 - \frac{v_+ - v_-}{v_+}w^2}{1 - w^2}}\,\mathrm{d}w\,,$$

then the standard substitution $w = \sin\varphi$ transforms it into the canonical form $E(m)$ (3.95) with $m = \frac{v_+ - v_-}{v_+}$,

$$J_{(v)} = 2\bar{\rho}\sqrt{2v_+}\,E(\tfrac{v_+ - v_-}{v_+})\,.$$

Accordingly, the time derivative $\partial_t C(t, K)$ takes the form

$$\partial_t C(t, K) = \frac{2V_0\,\mathrm{e}^{-t/8}}{(2\pi t)^{3/2}}\int_{s_0}^{\infty} \mathrm{d}u\, u\,\mathrm{e}^{-\frac{u^2}{2t}}\sqrt{v_+}E(\tfrac{v_+ - v_-}{v_+})$$

(recall that roots $v_\pm$ (3.96) are functions of u).

The last step is integrating with respect to time which may precede that with respect to u . The option time value becomes

$$\mathcal{O}(T, K) = \frac{2V_0}{(2\pi)^{3/2}}\int_{s_0}^{\infty} \mathrm{d}u\, u\,\sqrt{v_+}E(\tfrac{v_+ - v_-}{v_+})\int_0^T \mathrm{d}t\, t^{-\frac{3}{2}}\,\mathrm{e}^{-\frac{t}{8} - \frac{u^2}{2t}}\,.$$

Keeping only necessary factors consider time integral

$$J_{(t)} = \frac{u}{\sqrt{2\pi}}\int_0^T \mathrm{d}t\, t^{-\frac{3}{2}}\,\mathrm{e}^{-\frac{t}{8} - \frac{u^2}{2t}} \tag{3.98}$$

(u enters here as a parameter). Notice that the exponent involved can be rewritten in two following forms

$$-\frac{t}{8} - \frac{u^2}{2t} = -\frac{1}{2}\left(\frac{u}{\sqrt{t}} + \frac{\sqrt{t}}{2}\right)^2 + \frac{u}{2} = -\frac{1}{2}\left(-\frac{u}{\sqrt{t}} + \frac{\sqrt{t}}{2}\right)^2 - \frac{u}{2}$$

and introduce two variables

$$\xi_\pm(t) = -\frac{u}{\sqrt{t}} \pm \frac{\sqrt{t}}{2}\,,$$

each turning into $(-\infty)$ at $t = 0$. Then two forms of the exponential become

$$-\frac{t}{8} - \frac{u^2}{2t} = -\frac{\xi_-^2(t)}{2} + \frac{u}{2} = -\frac{\xi_+^2(t)}{2} - \frac{u}{2}.$$

Next, take power factors with differential

$$u\, t^{-\frac{3}{2}}\, \mathrm{d}t = -2u\, \mathrm{d}(t^{-\frac{1}{2}}) = \mathrm{d}\xi_+ + \mathrm{d}\xi_-\,,$$

so that integral $J_{(t)}$ can be transformed into

$$\begin{aligned} J_{(t)} &= \frac{\mathrm{e}^{\frac{u}{2}}}{\sqrt{2\pi}} \int_{-\infty}^{x_-} \mathrm{e}^{-\frac{\xi_-^2}{2}} \mathrm{d}\xi_- + \frac{\mathrm{e}^{-\frac{u}{2}}}{\sqrt{2\pi}} \int_{-\infty}^{x_+} \mathrm{e}^{-\frac{\xi_+^2}{2}} \mathrm{d}\xi_+ \\ &= \mathrm{e}^{\frac{u}{2}} \Phi\,(x_-) + \mathrm{e}^{-\frac{u}{2}} \Phi\,(x_+) \qquad x_\pm = -\frac{u}{\sqrt{T}} \pm \frac{\sqrt{T}}{2}\,. \end{aligned} \tag{3.99}$$

Returning to notation $T = t$, we come up with the following alternative expression for the option time value

$$\mathcal{O}(t, K) = \frac{V_0}{\pi} \int_{s_0}^{\infty} \mathrm{d}u \left[\mathrm{e}^{\frac{u}{2}} \Phi\left(-\tfrac{u+t/2}{\sqrt{t}}\right) + \mathrm{e}^{-\frac{u}{2}} \Phi\left(-\tfrac{u-t/2}{\sqrt{t}}\right) \right] \sqrt{v_+} E\left(\tfrac{v_+ - v_-}{v_+}\right),$$

with s_0 defined by (3.90) and $v_\pm(u, k)$ given in (3.96).

3.6 Option Value for the Log-Normal SABR

We provide derivation of the option value for the log-normal case with zero correlation using PDF for the LN SABR (3.22) and linking it to free 'normal' PDF $G^{(2)}$ (3.15). We start with the forward Kolmogorov equation for PDF $p(F, V)$ of log-normal F, V process $\mathrm{d}F_t = F_t V_t \mathrm{d}W_t^1, \quad \mathrm{d}V_t = V_t \mathrm{d}W_t$,

$$\partial_t p(F, V) = \frac{1}{2} \left(\partial_F^2 F^2 V^2 + \partial_V^2 V^2 \right) p(F, V) \tag{3.100}$$

and integrate it with payoff $(F - K)^+$, to get the option time derivative

$$\partial_t C(t, K) = \frac{1}{2} K^2 \int p(K, V) V^2 \mathrm{d}V.$$

PDF-s $p(F, V)$ and p_{inv} are related as $p(F, V) = (F V^2)^{-1}\, p_{\mathrm{inv}}$ (3.51), in turn p_{inv} was presented in the form $p_{\mathrm{inv}} = \mathrm{e}^{-\frac{x}{2}}\, \tilde{p}(t, x, y)$ with $x = \log \frac{F}{F_0}$ at $\rho = 0$, $\mathrm{e}^{-\frac{x}{2}} = \sqrt{\frac{F_0}{F}}$, $y = V$ and function $\tilde{p}(t, x, y)$ found in Sect. 3.3. Applying these relations, with

$F = K$, rewrite the option derivative as

$$\partial_t C(K, V) = \frac{\sqrt{K F_0}}{2} \int \tilde{p}(t, x, V)\, \mathrm{d}V \quad x = \log \frac{K}{F_0}\,.$$

Next, we link function $\tilde{p}$ to McKean kernel $G^{(2)}$. Compare Eqs. (3.43) and (3.76) satisfied by two and take Fourier transform in x of both, —with wave number k for $\tilde{p}$ and with k for $G^{(2)}$. Fourier images solve equations

$$\partial_t \hat{p} = \frac{y^2}{2}(\partial_y^2 - k^2 - \tfrac{1}{4})\hat{p}\,,$$
$$\partial_t \hat{G}^{(2)} = \frac{y^2}{2}(\partial_y^2 - k'^2)\hat{G}^{(2)}$$

and share the same boundary conditions and initial condition

$$\hat{p}|_{t=0} = \hat{G}^{(2)}|_{t=0} = y_0^2\, \delta(y - y_0).$$

It is evident that $\hat{p}$ and $\hat{G}^{(2)}$ coincide if $k'^2 = k^2 + 1/4$. This allows to express $\tilde{p}$ and then $\partial_t C$ in terms of $G^{(2)}$

$$\partial_t C(K, V) = \frac{\sqrt{K F_0}}{4\pi} \int \mathrm{e}^{\mathrm{i}kx} \mathrm{d}k \int \mathrm{e}^{-\mathrm{i}k'x'} \mathrm{d}x' \int G^{(2)}(t, s(x', V))\, \mathrm{d}V, \qquad (3.101)$$
$$k' = (k^2 + 1/4)^{1/2}\,,$$

The inner integral over V generates (with factor 2) the time derivative of the free normal call $\partial_t C_N(t, x')$, as seen from expression (3.86). Integrating over time we obtain relation between the log-normal and free normal time values, the latter is given by (3.92) with $s_0(x') = \sinh^{-1}(|x'|/V_0)$,

$$\mathcal{O}_{LN}(t, s_0) = \frac{\sqrt{K F_0}}{2\pi} \int \mathrm{e}^{\mathrm{i}kx}\, \mathrm{d}k \int \mathrm{e}^{-\mathrm{i}k'x'} \mathcal{O}_N(t, s_0(x'))\, \mathrm{d}x'\,.$$

Now substitute (3.92) for $\mathcal{O}_N(t, s_0(x'))$ and change the order of integration according to $|x'|/V_0 = \sinh s_0(x') < \sinh s$

$$\mathcal{O}_{LN}(t, s_0) = \frac{\sqrt{K F_0}}{\pi} \int_0^\infty \frac{G(t, s)}{\sinh s} \mathrm{d}s \int \mathrm{e}^{\mathrm{i}kx} \frac{\mathrm{d}k}{2\pi} \int_{-r_0}^{r_0} \mathrm{e}^{-\mathrm{i}k'x'} \sqrt{r_0^2 - x'^2}\, \mathrm{d}x'\,, \qquad (3.102)$$

here temporary notation $r_0 = V_0 \sinh s$ is used. The inner integral over x' is taken by substituting $x' = r_0 \cos\varphi$, integrating by parts, and recognizing the integral representation of Bessel function J_1,

$$I_{(x')} = r_0^2 \int_0^{\pi} e^{-ik' r_0 \cos\varphi} \sin^2\varphi \, \mathrm{d}\varphi = \frac{\pi r_0}{k'} J_1(k' r_0).$$

Next is the integral over k with $k' = \sqrt{k^2 + 1/4}$, which is equal to

$$I_{(k)} = r_0 \int_0^{\infty} \frac{J_1\left(r_0\sqrt{k^2+1/4}\right)}{\sqrt{k^2+1/4}} \cos kx \, \mathrm{d}k = 2 \sin\left(\frac{1}{2}\sqrt{r_0^2 - x^2}\right)$$

if $r_0 > |x|$, and zero otherwise. It can be computed by 'playing' with the integral representation of J_0 and differentiation with respect to a parameter, or found in [34], 6.677.

Plugging it into (3.102) and restoring $r_0 = V_0 \sinh s, \quad x = \log(K/F_0)$, we come up with the following expression for the time value of the 'log-normal' option

$$\mathcal{O}(t, K) = \frac{2\sqrt{KF_0}}{\pi} \int_{s_-}^{\infty} \frac{G(t, s)}{\sinh s} \sin\left(\frac{V_0}{2}\sqrt{\sinh^2 s - \sinh^2 s_-}\right) \mathrm{d}s, \tag{3.103}$$

$$\sinh s_- = \frac{|\log(K/F_0)|}{V_0}. \tag{3.104}$$

3.7 The Zero Correlation Case

Below we present analytics for the call option values with $\rho = 0$ and power β in the interval $0 < \beta < 1$, including asymptotics for large strikes and the sensitive ATM case with strikes K, close to spot F_0.

The case was initiated in the previous Chapter, where the main formula (2.72) for the option value was obtained by utilizing the explicit form (2.70) of the moment generating function $M_{\tau^{-1}}$ (MGF) of the inverse random time τ. The important formula (2.70) was only announced, with derivation postponed until now.

3.7.1 *Moment Generating Function of τ^{-1}*

Function

$$M_{\tau^{-1}}(\lambda) := \mathbb{E}\left[\exp\left(-\frac{\lambda}{\tau}\right)\right]$$

is computed with the help of the joint PDF of V and τ (3.6) as

$$\mathbb{E}\left[\exp\left(-\frac{\lambda}{\tau}\right)\right] = \int_0^{\infty} \mathrm{e}^{-\lambda/\tau} p(t; V, \tau \mid V_0) \, \mathrm{d}V \, \mathrm{d}\tau\,.$$

We substitute p (3.6) with function θ from (3.7) into the last integral. With the goal of first integrating over τ, we have to improve convergence of τ-integral, which can be achieved by integrating expression (3.7) for θ by parts, thus presenting function $\theta(r,t)$ in a different, yet equivalent form

$$\theta(r,t) = \frac{r}{\mathrm{i}2\pi\sqrt{2\pi t}} \int_{C_u} \exp\left(r\cosh u - \frac{u^2}{2t}\right) \sinh u \, \mathrm{d}u \, .$$

As a result, MGF $M_{\tau^{-1}}$ becomes

$$\mathbb{E}\left[\exp\left(-\frac{\lambda}{\tau}\right)\right] = \frac{\mathrm{e}^{-\frac{t}{8}}}{\sqrt{2\pi t}} \int_0^\infty \frac{\mathrm{d}V}{V} \frac{\mathrm{d}\tau}{\tau^2} \left(\frac{V}{V_0}\right)^{-1/2} \mathrm{e}^{-\frac{V^2+V_0^2+2\lambda}{2\tau}} V V_0$$
$$\frac{1}{2\pi i} \int_{C_u} \exp\left(\frac{V V_0}{\tau} \cosh u - \frac{u^2}{2t}\right) \sinh u \, \mathrm{d}u \, .$$

Performing first integration over τ, we gather all τ-dependent terms (including for convenience also factor $V V_0$)

$$\begin{aligned} I_{(\tau)} &= V V_0 \int_0^\infty \frac{\mathrm{d}\tau}{\tau^2} \exp\left(-\frac{V^2+V_0^2+2\lambda-2VV_0\cosh u}{2\tau}\right) \\ &= \left(\frac{V^2+V_0^2+2\lambda}{2VV_0} - \cosh u\right)^{-1} . \end{aligned}$$

Now denote

$$u_0(V,\lambda) = \cosh^{-1}\left(\frac{V^2+V_0^2+2\lambda}{2VV_0}\right) \tag{3.105}$$

and present $I_{(\tau)} = (\cosh u_0 - \cosh u)^{-1}$. The sought expectation takes the form

$$\mathbb{E}\left[\exp\left(-\frac{\lambda}{\tau}\right)\right] = \frac{\mathrm{e}^{-\frac{t}{8}}}{\sqrt{2\pi t}} \int_0^\infty \frac{\mathrm{d}V}{V} \left(\frac{V}{V_0}\right)^{-1/2} \frac{1}{2\pi\mathrm{i}} \int_{C_u} \frac{\mathrm{e}^{-\frac{u^2}{2t}} \sinh u}{\cosh u_0 - \cosh u} \mathrm{d}u \, .$$

Next, the integral over u is readily obtained with the help of the residue theory. Since the integrand decreases fast at $\Re u \to +\infty$, ends of contour C_u may be connected at $\Re u = +\infty$. In the horizontal strip surrounded by (now closed) C_u, the only singularity of the integrand, as a function of complex variable u, is the simple pole at $u = u_0$. Then according to Cauchy residue theorem

$$\frac{1}{2\pi\mathrm{i}} \int_{C_u} \frac{\mathrm{e}^{-\frac{u^2}{2t}} \sinh u}{\cosh u_0 - \cosh u} \mathrm{d}u = \mathrm{e}^{-\frac{u_0^2}{2t}} ,$$

leading to

$$\mathbb{E}\left[\exp\left(-\frac{\lambda}{\tau}\right)\right] = \frac{e^{-\frac{t}{8}}}{\sqrt{2\pi t}}\int_0^\infty \frac{dV}{V}\left(\frac{V}{V_0}\right)^{-1/2}\exp\left(-\frac{u_0^2(V,\lambda)}{2t}\right). \tag{3.106}$$

Turn now to the definition of u_0 (3.105) and introduce the amplitude

$$D^2 = \frac{V_0^2 + 2\lambda}{V_0^2},$$

so that

$$\cosh u_0 = D\left(\frac{V}{2DV_0} + \frac{DV_0}{2V}\right).$$

Now, parametrize

$$\frac{V}{DV_0} = e^w,$$

resulting in

$$\cosh u_0(w) = D\cosh w \tag{3.107}$$

The expectation takes the form

$$\mathbb{E}\left[\exp\left(-\frac{\lambda}{\tau}\right)\right] = \frac{e^{-\frac{t}{8}}}{\sqrt{D}\sqrt{2\pi t}}\int_{-\infty}^{\infty} dw\, e^{-w/2}\, e^{-\frac{u_0(w)^2}{2t}}.$$

Using symmetry property of $u_0(w)$, keep only the even part of the exponential $e^{-w/2} \to \cosh\frac{w}{2} \to 2\cosh\frac{w}{2}\mathbf{1}_{w>0}$, use also $\cosh\frac{w}{2}\,dw = 2d\left(\sinh\frac{w}{2}\right)$, then

$$\mathbb{E}\left[\exp\left(-\frac{\lambda}{\tau}\right)\right] = \frac{4e^{-\frac{t}{8}}}{\sqrt{D}\sqrt{2\pi t}}\int_{w=0}^{\infty} d\left(\sinh\frac{w}{2}\right) e^{-\frac{u_0(w)^2}{2t}}.$$

Now, change variable of integration $w \to u = u_0(w)$, express

$$\sinh\frac{w}{2} = \sqrt{\frac{\cosh w - 1}{2}} = \frac{\sqrt{\cosh u - D}}{\sqrt{2D}},$$

and integrate by parts, thus coming up with

$$\mathbb{E}\left[\exp\left(-\frac{\lambda}{\tau}\right)\right] = \frac{2e^{-\frac{t}{8}}}{Dt\sqrt{\pi t}}\int_{\cosh^{-1}(D)}^{\infty} e^{-\frac{u^2}{2t}}\sqrt{\cosh u - D}\, u\, du,$$

where the lower limit follows from (3.107) at $w = 0$. Finally, denote $\cosh^{-1}(D) = s$, implying that

$$\cosh^2 s = D^2 = 1 + \frac{2\lambda}{V_0^2}, \text{ or } \sinh^2 s = \frac{2\lambda}{V_0^2},$$

and the last expression transforms into

$$\mathbb{E}\left[\exp\left(-\frac{\lambda}{\tau}\right)\right] = \frac{G(t,s)}{\cosh s}, \qquad \sinh s = \frac{\sqrt{2\lambda}}{V_0}, \tag{3.108}$$

where $G(t,s)$ is exactly CDF, given by (3.82).

3.7.2 Marginal Distribution of the Underlying *F*

We will work with PDF $p_q(q)$ of variable $q = \frac{F^{1-\beta}}{1-\beta}$. PDF of F is then restored as $P_F(F) = |\frac{\mathrm{d}q}{\mathrm{d}F}|\, p_q(q(F))$. In turn, p_q is related to PDF of $x = q^2$ as $p_q(q) = 2q\, p_x(q^2)$, and p_x, conditioned on a given random time τ, is given by (2.26) (absorbing), or (2.25) (reflecting) from Chap. 2 with τ standing for t, thus

$$p_{a/r}^{(-|\nu|)}(q \mid \tau, q_0) = \frac{q}{\tau}\left(\frac{q}{q_0}\right)^{-|\nu|} \mathrm{e}^{-\frac{q^2+q_0^2}{2\tau}}\, I_{\pm|\nu|}\left(\frac{q q_0}{\tau}\right). \tag{3.109}$$

PDF of q in the actual time t is then obtained as expectation of the last expression over random time τ. Using integral representation (2.57)

$$I_\nu(r) = \frac{1}{2\pi \mathrm{i}} \int_{C_u} \mathrm{e}^{r\cosh u - \nu u}\, \mathrm{d}u$$

with the same Π-shaped contour (2.58) and denoting temporarily

$$\lambda(q,w) = \tfrac{1}{2}(q^2 + q_0^2 - 2qq_0 \cosh u),$$

we get

$$p(t;q \mid q_0) = q\left(\frac{q}{q_0}\right)^{-|\nu|} \frac{1}{2\pi \mathrm{i}} \int_{C_u} \mathrm{e}^{\mp|\nu|u}\, \mathbb{E}\left[\frac{1}{\tau}\mathrm{e}^{-\lambda/\tau}\right] \mathrm{d}u.$$

The expected value can be expressed as

$$\mathbb{E}\left[\frac{1}{\tau}\mathrm{e}^{-\lambda/\tau}\right] = -\partial_\lambda\, \mathbb{E}\left[e^{-\lambda/\tau}\right]$$

that allows to use directly result (3.108) with

$$\sinh^2 s = \frac{2\lambda(q,w)}{V_0^2} = \frac{q^2 + q_0^2 - 2qq_0\cosh u}{V_0^2}. \tag{3.110}$$

Passing from derivative ∂_λ to ∂_s, we find

$$p(t;q\mid q_0)=\frac{q}{V_0^2}\left(\frac{q}{q_0}\right)^{-|\nu|}\int_{C_u}\frac{\mathrm{d}u}{2\pi\mathrm{i}}\mathrm{e}^{\mp|\nu|u}\,\frac{1}{\cosh s\,\sinh s}\frac{\partial}{\partial s}\left(-\frac{G(t,s)}{\cosh s}\right).$$

Finally, develop three legs of contour C_u; on the vertical leg, $(-\mathrm{i}\pi,\mathrm{i}\pi)$, $u=\mathrm{i}\phi$ and $\cosh u=\cos\phi$ in (3.110), and on the horizontal legs $u=\psi\pm\mathrm{i}\pi$ and $\cosh u=-\cosh\psi$ in (3.110). We obtain PDF of q in the form

$$\begin{aligned}p(t;q\mid q_0)=\frac{q}{\pi V_0^2}\left(\frac{q}{q_0}\right)^{-|\nu|}&\left\{\int_0^{\pi}\mathrm{d}\phi\,\frac{\cos(\nu\phi)}{\cosh s\,\sinh s}\frac{\partial}{\partial s}\left(-\frac{G(t,s)}{\cosh s}\right)\right.\\&\left.\mp\sin(|\nu|\pi)\int_0^{\infty}\mathrm{d}\psi\,\mathrm{e}^{\mp|\nu|\psi(s)}\frac{1}{\cosh s\,\sinh s}\frac{\partial}{\partial s}\left(-\frac{G(t,s)}{\cosh s}\right)\right\}\end{aligned}\tag{3.111}$$

(it is worth noticing that $\partial s\,(-G/\cosh s)$ is always positive).

Small F (hence small q). It is difficult to extract small q dependence from the last expression. The reason is that in the integral representation of the Bessel function, expansion at small arguments is hidden. Instead we try to take the expected value (over τ) of the leading term in the conditional PDF (3.109). The leading terms at small q in the absorbing and in reflecting PDF are

$$p_a^{(-|\nu|)}(q\mid\tau,q_0)\approx\frac{1}{\Gamma(1+|\nu|)}\frac{q}{\tau}\left(\frac{q_0^2}{2\tau}\right)^{|\nu|}\exp\left\{-\frac{q_0^2}{2\tau}\right\},$$

$$p_r^{(-|\nu|)}(q\mid\tau,q_0)\approx\frac{1}{\Gamma(1-|\nu|)}\frac{q}{\tau}\left(\frac{2\tau}{q^2}\right)^{|\nu|}\exp\left\{-\frac{q_0^2}{2\tau}\right\}.$$

It is essential that expected values over τ exist in both cases (integrals with the joint PDF $p(V,\tau)$ are converging), thus making such expansion legitimate. We can state that the marginal PDF of the SABR process F_t inherits at small F behavior of the conditional PDF of the underlying CEV process F_τ. Regarding small F, we recall that the continuous part of the absorbing CEV PDF has a norm defect, with a finite probability of $F=0$, called also mass at zero (2.32). The SABR process inherits this property too, with the 'SABR' zero mass to be obtained as the expectation of the 'CEV' one (2.32),

$$P_{\mathrm{SABR}}(F_t=0)=\frac{1}{\Gamma(|\nu|)}\mathbb{E}\left[\Gamma(|\nu|,q_0^2/2\tau)\right].$$

We refer to [24, 36, 37] for intensive study of the mass at zero.

Asymptotics at **large** F is considered later, alongside with the option values.

3.7.3 Option Time Value

We recall expression for the time option value from Chap. 2 (2.72)

$$\mathcal{O}(t, K, F_0) = \frac{1}{\pi}\sqrt{KF_0}\left\{\int_0^{\pi} d\phi \, \frac{\sin\phi \, \sin(|\nu|\,\phi)}{b-\cos\phi}\,\frac{G(t, s(\phi))}{\cosh s(\phi)} + \sin(|\nu|\,\pi)\int_0^{\infty} d\psi \, \frac{\sinh\psi}{b+\cosh\psi}\mathrm{e}^{\mp|\nu|\psi}\,\frac{G(t, s(\psi))}{\cosh s(\psi)}\right\}. \tag{3.112}$$

Parameters involved and relations between s, ϕ, and ψ are given in Chap. 2, (2.42, 2.73, 2.74, 2.56). Here we will use

$$q_K = \sqrt{x_K} = \frac{K^{1-\beta}}{1-\beta} \quad \text{and} \quad q_0 = \sqrt{x_0} = \frac{F_0^{1-\beta}}{1-\beta}, \tag{3.113}$$

so that

$$b = \frac{q_K^2 + q_0^2}{2q_K q_0}.$$

Above we have kept both signs in the exponential $e^{\mp|\nu|\psi}$ (absorbing/reflecting) keeping in mind further application to the mixture SABR model (Chap. 5). For the classical SABR, though, the coherent choice is absorbing ($e^{-|\nu|\psi}$).

One can then change the variables of integration to s (see [9] for more details) to obtain the option price in the following compact form [10]

$$\mathcal{O}(t, K, F_0) = \frac{2}{\pi}\sqrt{KF_0}\left\{\int_{s_-}^{s_+} ds \frac{\sin(|\nu|\phi(s))}{\sinh s} G(t, s) + \sin(|\nu|\pi)\int_{s_+}^{\infty} ds \frac{\mathrm{e}^{\mp|\nu|\psi(s)}}{\sinh s} G(t, s)\right\}. \tag{3.114}$$

Now parameters ϕ and ψ are treated as functions of s,

$$\phi(s) = 2\tan^{-1}\sqrt{\frac{\sinh^2 s - \sinh^2 s_-}{\sinh^2 s_+ - \sinh^2 s}}, \tag{3.115}$$

$$\psi(s) = 2\tanh^{-1}\sqrt{\frac{\sinh^2 s - \sinh^2 s_+}{\sinh^2 s - \sinh^2 s_-}}, \tag{3.116}$$

and the new limits of integration are found after careful rearrangement

$$s_- = \sinh^{-1}\left(\frac{|q_K - q_0|}{V_0}\right) \tag{3.117}$$

$$s_+ = \sinh^{-1}\left(\frac{q_K + q_0}{V_0}\right). \tag{3.118}$$

Note that the option price depends[5] on the parameters q_0, q_K and V_0 through dimensionless s_- and s_+.

The integration can be performed numerically in an efficient manner – the integrands are smooth functions of the parameters. There is also a good approximation for the function G [10]

$$G(t, s) \simeq \sqrt{\frac{\sinh s}{s}}\, e^{-\frac{s^2}{2t}-\frac{t}{8}} \left(R(t, s) + \delta R(t, s)\right), \tag{3.119}$$

where

$$\begin{aligned} R(t, s) &= 1 + \frac{3\,t\,g(s)}{8\,s^2} - \frac{5t^2\left(-8s^2 + 3g^2(s) + 24g(s)\right)}{128\,s^4} \\ &+ \frac{35t^3\left(-40s^2 + 3g^3(s) + 24g^2(s) + 120g(s)\right)}{1024\,s^6}, \\ g(s) &= s \coth s - 1, \end{aligned} \tag{3.120}$$

and the correction $\delta R(t, s)$ is defined as

$$\delta R(t, s) = e^{\frac{t}{8}} - \frac{3072 + 384t + 24t^2 + t^3}{3072} \tag{3.121}$$

to guarantee that $G(t, 0) = 1$. This is an effective small time expansion.

3.7.4 N and Log-N SABR Option Values as Limits of General Case

We reproduce original expressions (3.93) and (3.103) by taking limits of the general case (3.114) at β tending to 0 and 1.

Normal SABR ($\beta = 0$). In this case $|\nu| = \frac{1}{2(1-\beta)} = \frac{1}{2}$, $q = F$, $q_K = K$, and all we need is to compute functions $\sin \frac{\phi}{2}$ and $\exp(\mp\frac{\psi}{2})$. Denote for brevity

$$R_\pm = \sqrt{\sinh^2 s - \sinh^2 s_\pm}$$

with $s_\pm$ given by (3.118, 3.117), note also that $R_-^2 - R_+^2 = \frac{4KF_0}{V_0^2}$ and that R_+ is used only for $s > s_+$. Then, according to definitions (3.115–3.117), we get

[5] Except the square root $\sqrt{KF_0}$.

$$\sin\frac{\phi}{2} = \frac{\tan\frac{\phi}{2}}{\sqrt{1+\tan^2\frac{\phi}{2}}} = \frac{R_-}{2\sqrt{KF_0}}V_0\,,$$

$$e^{\mp\frac{\psi}{2}} = \frac{1\mp\tanh\frac{\psi}{2}}{\sqrt{1-\tanh^2\frac{\phi}{2}}} = \frac{R_-\mp R_+}{2\sqrt{KF_0}}V_0\,,$$

resulting in the following expression

$$\begin{aligned} C(t,K,F_0)-(F_0-K)^+ &= \frac{V_0}{\pi}\left\{\int_{s_-}^{s_+}\mathrm{d}s\,\frac{G(t,s)}{\sinh s}R_- + \int_{s_+}^{\infty}\mathrm{d}s\,\frac{G(t,s)}{\sinh s}(R_-\mp R_+)\right\} \\ &= \frac{V_0}{\pi}\left\{\int_{s_-}^{\infty}\mathrm{d}s\frac{G(t,s)}{\sinh s}R_- \mp \int_{s_+}^{\infty}\mathrm{d}s\frac{G(t,s)}{\sinh s}R_+\right\}. \end{aligned}$$

Thus, the obtained time value of the call option is in line with our earlier results,

$$C(t,K,F_0)-(F_0-K)^+ = \mathcal{O}(t,s_-)\mp\mathcal{O}(t,s_+)\,, \tag{3.122}$$

where $\mathcal{O}(t,s)$ is exactly the free normal time value (3.92),

$$\mathcal{O}(t,s_0) = \frac{V_0}{\pi}\int_{s_0}^{\infty} Ds\,\frac{G(t,s)}{\sinh s}\sqrt{\sinh^2 s-\sinh^2 s_0}\,. \tag{3.123}$$

The upper (lower) sign refers to the absorbing (reflecting) boundary.

Log-normal SABR ($\beta = 1$). This case requires an accurate transition to the limit $\beta\to 1$. First of all notice that $q(F)$ behaves as

$$q = \frac{F^{1-\beta}}{1-\beta} \approx \frac{1}{1-\beta} + \log F\,,$$

meaning that while both q_K and q_0 tend to infinity, their difference remains finite so that

$$\sinh s_- = \frac{|q_K-q_0|}{V_0} \to \frac{|\log\frac{K}{F_0}|}{V_0}\,,$$

$$\sinh s_+ = \frac{q_K+q_0}{V_0} \approx \frac{2}{(1-\beta)V_0} \to \infty\,.$$

This implies in turn that the second integral in (3.114) tends to zero. In the first integral, function $\sin(|\nu|\phi(s))$ is oscillating fast which means that only small angles $\phi(s)$ do contribute. Turning to definition of ϕ (3.115) we see that at moderate s and large s_+ the leading order of $\phi(s)$ looks like

$$\phi(s) \approx 2\,\frac{\sqrt{\sinh^2 s - \sinh^2 s_-}}{\sinh s_+} \approx (1-\beta)V_0\sqrt{\sinh^2 s - \sinh^2 s_-}\,,$$

leading to

$$|\nu|\phi(s) = \frac{\phi(s)}{2(1-\beta)} \to \frac{V_0}{2}\sqrt{\sinh^2 s - \sinh^2 s_-}\,.$$

We come up with the time value of call (3.103)

$$C(t,K,F_0) - (F_0-K)^+ = \frac{2}{\pi}\sqrt{KF_0}\int_{s_-}^{\infty} \mathrm{d}s\,\frac{G(t,s)}{\sinh s}\sin\left(\frac{V_0}{2}\sqrt{\sinh^2 s - \sinh^2 s_-}\right). \tag{3.124}$$

3.7.5 Large Strike Asymptotics

In derivation that follows we try to use asymptotics of the CEV PDF (3.109). We start with presenting the SABR call value $C(t,K)$ (for $\rho = 0$) as the expectation over τ of its CEV counterpart

$$C(t,K) = \mathbb{E}\left[C_{\mathrm{CEV}}(\tau,K)\right] = \int C_{\mathrm{CEV}}(\tau,K)\,p(t;V,\tau)\,\mathrm{d}V\,\mathrm{d}\tau\,.$$

Turn now to expression (2.53) for the CEV option time derivative, which was obtained in Chap. 2 by integrating the CEV FK equation (2.52) with payoff $(F-K)^+$ and then to (2.54), where CEV PDF was specified. With random time τ standing for t and in terms of $q = x^{1/2}$, they take the form

$$\partial_\tau C_{\mathrm{CEV}}(\tau,K) = \frac{F^{2\beta}}{2}\,p_{\mathrm{CEV}}(\tau,K) \tag{3.125}$$

$$= |\nu|\frac{\sqrt{KF_0}}{\tau}e^{-\frac{q_K^2+q_0^2}{2\tau}}I_{\pm|\nu|}\left(\frac{q_K q_0}{\tau}\right). \tag{3.126}$$

If we applied the integral representation for the Bessel function, we could recover the general expression (3.112). Being interested, though, in large strikes, we take instead the asymptotics of the Bessel function for large arguments (the leading term) $I_\mu(r) \simeq \frac{\mathrm{e}^r}{\sqrt{2\pi r}}$ and find

$$\partial_\tau C_{\mathrm{CEV}}(\tau,K) \simeq 2|\nu|\sqrt{\frac{KF_0}{q_K q_0}}\left\{\frac{\mathrm{e}^{-(q_K-q_0)^2/2\tau}}{2\sqrt{2\pi\tau}}\right\}. \tag{3.127}$$

Now look at the general relation (3.125), taken at $\beta = 0$, when the the CEV process becomes a Brownian motion. In braces we have the Gaussian PDF for the free normal variable q_K, therefore $\{\cdot\} = \partial_\tau C_{\text{freeN}}(\tau, q_K)$. Using this in (3.127) and taking into account that the factor outside braces can be rewritten as $(K F_0)^{\beta/2}$, we find

$$C_{\text{CEV}}(\tau, K) \simeq (K F_0)^{\beta/2}\, C_{\text{freeN}}(\tau, q_K)\,.$$

Pay attention that on the left side we have the real strike K, while on the right it is q_K standing for strike. Finally, taking expectation over τ, we obtain

$$\mathcal{O}_{\text{SABR}}(t, K) \simeq (K F_0)^{\beta/2} \mathcal{O}_N(t, s_K)\,, \tag{3.128}$$

where $\mathcal{O}_N(t, s_K)$ is the free normal SABR time value (3.123) with

$$\sinh^2 s_K = \frac{(q_k - q_0)^2}{V_0^2}\,.$$

To complete derivation we need asymptotics of the normal time value $\mathcal{O}_N(t, s_K)$ (3.123) for large s_0. Using asymptotics (3.84) of CDF $G(t, s)$ and making in (3.123) substitution

$$s = (s_0^2 + w^2)^{1/2} \simeq s_0 + \frac{w^2}{2s_0}, \qquad \mathrm{d}s = \frac{w\, \mathrm{d}w}{s_0}\,,$$

we keep only lowest order terms in w to get the leading order of asymptotics. In fact, the only other term that needs expansion is

$$\left(\sinh^2 s - \sinh^2 s_0\right)^{1/2} \simeq \left(2 \sinh s_0 \cosh s_0 \frac{w^2}{2s_0}\right)^{1/2} = \sqrt{\frac{\sinh s_0 \cosh s_0}{s_0}}\, w\,,$$

factors $(\sinh s/s)^{1/2}$ in G and $\sinh s$ in the denominator may be taken at $s = s_0$. As a result,

$$\mathcal{O}_N(t, s_0) \simeq \frac{V_0}{\sqrt{2\pi}} (\cosh s_0)^{1/2}\, t^{3/2}\, s_0^{-2} \exp\left(-\frac{t}{8} - \frac{s_0^2}{2t}\right).$$

Using this in (3.128) with $s_0 = s_K$ and

$$\cosh s_K \simeq \sinh s_K \simeq \frac{1}{2}\, \mathrm{e}^{s_K} \simeq \frac{q_k}{V_0}, \qquad s_K \simeq \log \frac{2q_K}{V_0}\,,$$

we finally come up with the asymptotics of the **SABR option** value

$$\mathcal{O}_{\text{SABR}}(t, K) \simeq K^{1/2} F_0^{\beta/2} \sqrt{\frac{V_0}{2\pi(1-\beta)} \frac{t^{3/2}}{s_K^2}} \exp\left(-\frac{t}{8} - \frac{s_K^2}{2t}\right).$$

Then the asymptotics of the **marginal PDF** $p(t, F)$ at $F = K$ follows as

$$p(t, K) = \frac{\partial^2 \mathcal{O}_{\mathrm{SABR}}(t, K)}{\partial K^2}.$$

The main contribution into derivative arises from the exponential

$$\frac{\partial^2}{\partial K^2} \exp\left(-\frac{s_K^2}{2t}\right) \simeq \left(\frac{s_K}{t}\frac{\partial s_K}{\partial K}\right)^2 \exp\left(-\frac{s_K^2}{2t}\right), \qquad \frac{\partial s_K}{\partial K} \simeq \frac{\partial \log q_K}{\partial K} = \frac{1-\beta}{K}.$$

The result looks like

$$p(t, K) \simeq (1-\beta)^{3/2}\, V_0^{1/2}\, K^{-3/2}\, F_0^{\beta/2}\, \frac{\exp\left(-\frac{t}{8} - \frac{s_K^2}{2t}\right)}{\sqrt{2\pi t}}.$$

3.7.6 SABR Option Value, Close to ATM

We consider a sensitive case of strike K, close to spot F_0. We obtain expansion of the full option value (the time value $\mathcal{O}(t, K, F_0)$ (3.114) plus the intrinsic value $(F_0 - K)^+$) up to the second order of $(K - F_0)$. It is impossible to present the whole volume of calculations here. We describe the method and show the results.

Expansion of the first integral in (3.114) in small s_- poses the main difficulty, because angle $\phi(s)$ (3.115) changes in a wide range from 0 to π, that does not allow to expand it directly in small s_-. This is an example of a multiscale problem and can be treated by the separation of scales. Introduce an auxiliary parameter s_0, such that

$$s_- \ll s_0 \ll s_+,$$

then split integral as $\int_{s_-}^{s_+}(\cdot)\mathrm{d}s = \left(\int_{s_-}^{s_0} + \int_{s_0}^{s_+}\right)(\cdot)\mathrm{d}s$.

In the interval $s_- < s < s_0$, the whole integrand may be expanded at small s and s_-, note though, that in expansion of $\sin(|\nu|\phi(s))$ with

$$\phi = 2 \arcsin\left(\frac{V_0^2}{4 q_K q_0}(\sinh^2 s - \sinh^2 s_-)\right)^{1/2}, \tag{3.129}$$

the square root must be kept intact. Perform integration.

In the complementary interval $s_0 < s < s_+$, expand $\sin(|\nu|\phi(s))$ at small s_-. Still, formal expansion of the square root in (3.129) is not allowed, as it would violate condition $\phi(s_+) = \pi$, and a more elaborate expansion scheme was exploited. Arising integrals from s_0 to s_+ are checked for convergence at $s = 0$ and integrated by parts, if needed, to reach convergence. Then each (converging) integral is split as $\int_{s_0}^{s_+}(\cdot)\mathrm{d}s = \left(\int_0^{s_+} - \int_0^{s_0}\right)(\cdot)\mathrm{d}s$ and the latter is expanded at small s_0. The result is combined with the previous one for the interval $(s_- < s < s_0)$. If computations are performed accurately, all terms, involving a separation parameter s_0, will cancel one another. And they did annihilate. The rest, combined with the second integral in (3.114) (from s_+ to infinity), is treated as a function $f(s_+)$ of s_+. Recall that

$$\sinh s_+ = \frac{q_K + q_0}{V_0} = \frac{2q_0}{V_0} + \frac{q_K - q_0}{V_0},$$

introduce

$$S = \sinh^{-1} \frac{2q_0}{V_0}, \tag{3.130}$$

and expand $f(s_+)$ around S at small δq.

Finally, expand all factors, depending on K, like $(KF_0)^{1/2}$ and q_K, in terms of $(K - F_0)$.

Upon completing these steps, we have obtained the following expansion for the option value

$$C = \frac{2F_0}{\pi}\left[\int_0^S \frac{\sin(|\nu|\phi(s))}{\sinh s} G\,\mathrm{d}s + \sin(|\nu|\pi)\int_S^\infty \mathrm{e}^{\mp|\nu|\psi(s)} \frac{G}{\sinh s}\,\mathrm{d}s\right]$$

$$+ (K - F_0)\left\{-\frac{1}{2} + \frac{1}{\pi}\int_0^S \frac{\sin(|\nu| - 1/2)\phi(s)}{\cos\frac{\phi(s)}{2}} \frac{G}{\sinh s}\,\mathrm{d}s \right.$$

$$\left. \pm \frac{\sin(|\nu|\pi)}{\pi}\int_S^\infty \frac{\mathrm{e}^{\mp(|\nu|-1/2)\psi(s)}}{\sinh\frac{\psi(s)}{2}} \frac{G}{\sinh s}\,\mathrm{d}s\right\}$$

$$+ \frac{(K - F_0)^2\, F_0^{-\beta}}{2\pi V_0}\left\{\int_0^S \mathrm{d}s\, \frac{\cos(\nu\phi(s))}{\cos\frac{\phi(s)}{2}} \frac{1}{\sinh s}\frac{\partial}{\partial s}\left(-\frac{G}{\cosh s}\right)\right.$$

$$\left. \mp \sin(|\nu|\pi)\int_S^\infty \mathrm{d}s\, \frac{\mathrm{e}^{\mp|\nu|\psi(s)}}{\sinh\frac{\psi(s)}{2}} \frac{1}{\sinh s}\frac{\partial}{\partial s}\left(-\frac{G}{\cosh s}\right)\right\}.$$

Upper and lower signs correspond to absorbing and reflecting boundaries; S is given by (3.130), and $\phi(s)$, $\psi(s)$ are defined as

$$\phi(s) = 2\arcsin\left(\frac{\sinh s}{\sinh S}\right) \quad \text{for } s < S,$$

$$\psi(s) = 2\cosh^{-1}\left(\frac{\sinh s}{\sinh S}\right) \quad \text{for } s > S.$$

All integrals are convergent. In the 0-th and 1-st order terms integrals from 0 to S converge at $s = 0$ because $\phi(s) \sim s$ at small s. In the 2-nd order the first integral converges at $s = 0$ despite factor $(\sinh s)^{-1}$ because $G \sim s^2$ at small s. All integrands in the 1-st and 2-nd order have singularities at $s = S$ because at this point $\phi = \pi$ and $\psi = 0$, but these singularities are of the square root type and therefore integrable.

For the normal SABR, $\beta = 0,\ |\nu| = 1/2,\ q = F$, expansion simplifies to

$$C = C_0 \frac{V_0}{\pi} + C_1(F_0 - K) + C_2 \frac{(K - F_0)^2}{2\pi V_0}$$

with

$$\begin{aligned}
C_0 &= \int_0^\infty G(s)\, \mathrm{d}s - \int_S^\infty \frac{G(s)}{\sinh s} \sqrt{\sinh^2 s - (2F_0/V_0)^2}\, \mathrm{d}s\,, \\
C_1 &= \frac{1}{2} - \frac{2F_0}{V_0} \int_S^\infty \frac{G(s)}{\sinh s} \frac{\mathrm{d}s}{\sqrt{\sinh^2 s - (2F_0/V_0)^2}}\,, \\
C_2 &= \int_0^\infty \left(-\frac{\partial G}{\partial s} \right) \coth s\, \mathrm{d}s - \int_S^\infty \frac{\partial}{\partial s} \left(- \frac{G}{\cosh s}\right) \frac{\mathrm{d}s}{\sqrt{\sinh^2 s - (2F_0/V_0)^2}}\,, \\
S &= \sinh^{-1} \frac{2F_0}{V_0} \quad (= s_+ | K = F_0)\,.
\end{aligned}$$

The option value is a smooth function of strike K across the ATM point F_0.

3.8 Appendix A. Integrals with Square Root $\sqrt{\cosh s - \cosh u}$

On several occasions we face contour integrals of the type

$$J = \frac{1}{2\mathrm{i}} \int_{C_u} \frac{F(u)}{\sqrt{\cosh s - \cosh u}} \mathrm{d}u \qquad s > 0\,, \tag{3.131}$$

where the path of integration C_u in the complex plane u is defined like in (2.58), recall, it consists of three sides of the infinite rectangle with vertices at $\infty - \pi\mathrm{i},\ -\pi\mathrm{i},\ \pi\mathrm{i},\ \infty + \pi\mathrm{i}$, run in this order. It is supposed that function $F(u)$ is analytic in the half-strip Π, surrounded by the contour C_u, and is decreasing fast at $\Re u \to \infty$ (typically $F(u) \propto \mathrm{e}^{-\frac{u^2}{2t}}$). Function

$$f(u) = (\cosh s - \cosh u)^{1/2}\,, \tag{3.132}$$

defined originally on contour C_u, can be continued analytically into the half-strip Π, equipped with the cut along the real axis from the branching point $u = s$ to $+\infty$. The cut allows to select a single valued analytic branch of $f(u)$, fixed by the condition of $f(u)$ being positive on C_u, as discussed in Sect. 3.2.

According to the Cauchy theorem, the value of integral J of the analytic function $F(u)/f(u)$ depends only on the ends of the path C_u. Even more, the ends themselves may be shifted along the vertical line $\Re u = \infty$ due to decreasing $F(u)$. This means that without changing the value of integral J we may choose any path C that sustains the global properties of the original contour C_u; it starts at $\Re u = +\infty$ on or below the lower bank of the cut, terminates at $\Re u = +\infty$ on or above the upper bank, and goes around the cut in the negative direction (crossing cut is not allowed). This said, we squeeze contour C_u, placing integration directly onto the banks of the cut, plus an infinitesimal circumference surrounding point $u = s$. Contribution from the latter vanishes as its radius tends to zero, because the singularity $1/f(u) \propto (\delta u)^{-1/2}$ is integrable. This leaves only two integrals, taken in opposite directions along the banks.

We need now values of function $f(u)$ on the cut banks. On either bank, $u = \Re u > s$ so that $f(u)$ (3.132) is pure imaginary. Proper signs can be determined by moving from contour C_u to the cut banks. Take u on the far right end of the upper leg of C_u, $u = R + \mathrm{i}\pi, \quad R \gg s$, where asymptotics of $f(u)$ (3.132) is $\sqrt{2} f(u) \approx \mathrm{e}^{R/2} = -\mathrm{i}\mathrm{e}^{u/2}$. By continuity, this asymptotics remains valid, when moving down off C_u and reaching the upper bank at $u = R$ with $\sqrt{2} f(u) \approx -\mathrm{i}\mathrm{e}^{R/2}$. We conclude that on the upper bank

$$f(u) = -\mathrm{i}\,|f(u)| = -\mathrm{i}\sqrt{\cosh u - \cosh s}$$

In the same manner, moving upward from the far right end of the lower leg of C_u, we find that on the lower bank

$$f(u) = +\mathrm{i}\sqrt{\cosh u - \cosh s}$$

(We write $\sqrt{(\cdot)}$, rather than $(\cdot)^{1/2}$, to underline that the expression under square root is positive and that the positive value of the square root is taken).

Signs of $f(u)$ and directions of integration are opposite on the two banks, meaning that contributions from the two are equal, and the whole integral J is twice that along the upper bank,

$$J = \frac{1}{2\mathrm{i}} \int_{C_u} \frac{F(u)}{\sqrt{\cosh s - \cosh u}} \,\mathrm{d}u = \int_s^\infty \frac{F(u)}{\sqrt{\cosh u - \cosh s}} \,\mathrm{d}u \tag{3.133}$$

We use this transformation when computing PDFs of normal, log-normal, and general SABR with zero correlation; the moment generating function of the inverse random time τ; and option values. In the case of Log-normal SABR the integral to be transformed is slightly different from (3.131), but the consideration follows the same lines. The original and transformed integrals look like follows

$$\begin{aligned} \tilde{J} &= \frac{1}{2\mathrm{i}} \int_{C_u} \frac{F(u)\, E^{-\lambda\sqrt{\cosh s - \cosh u}}}{\sqrt{\cosh s - \cosh u}} \,\mathrm{d}u \\ &= \int_s^\infty \frac{F(u) \cos(\lambda\sqrt{\cosh u - \cosh s})}{\sqrt{\cosh u - \cosh s}} \,\mathrm{d}u \,. \end{aligned} \tag{3.134}$$

Chapter 4
Classic SABR Model: Heat Kernel Expansion and Projection on Solvable Models

4.1 Introduction

In the original article [39], Hagan et al. came up with an approximation formula for European option prices. However, the approximation quality rapidly degrades with time, for example, for maturities larger than 10Y the error in implied volatility can be 1% or more even for ATM values. Moreover, one can easily observe bad approximation behavior for extreme strikes which sometimes prevents obtaining a valid probability density function.

The original approach [39] was then formulated in terms of the widely used heat kernel (HK) [15, 42, 43, 67] expansion, which represents an asymptotic series in small times. We underline that this (diverging) expansion degrades with growing time, besides, the HK expansion does not 'feel' boundary conditions, thus generating errors at spots/strikes approaching zero. A more rigorous method beyond the HK expansion has been developed in [26, 46, 47] to examine delicate degeneracies of the SABR model at the origin, including analysis and a robust discretization scheme for small strikes and spots. More problems with the HK expansion have been addressed in a number of works; the upper bound on the time to expiry [14, 27], the impact of the mass at zero spot on existing implied volatility approximations [24, 36, 37], numerical simulation [58], to mention a few.

We do not cover these topics here, as our goal was to avoid those problems, rather than address them. Our usage of the HK expansion is concentrated in the ATM region, where it works fine, and is solely for the purpose of mapping onto the exactly solvable model, as described below.

In the present chapter we review results on the classical SABR model based on the heat kernel asymptotic expansion. We follow here Paulot's narrative with some modification. Then, we pass to the approximate results for SABR option pricing, combining asymptotic expansion close to ATM with mapping procedure (Antonov-Misirpashaev [7]) onto the exactly solvable zero correlation SABR model described in the previous chapter. Finally, we provide supporting numerical results. We start

A. Antonov et al., *Modern SABR Analytics*,
SpringerBriefs in Quantitative Finance,
https://doi.org/10.1007/978-3-030-10656-0_4

with discussion of general properties of absorbing and reflecting solutions for SABR PDF preceded by some preliminary remarks.

Recall that the classical SABR model

$$dF_t = F_t^{\beta}\, v_t\, dW_t^1\,, \tag{4.1}$$

$$dv_t = \gamma\, v_t\, dW_t^2 \tag{4.2}$$

uses a widely known set of five parameters $\{F_0, v_0, \beta, \gamma, \rho\}$. Usually, the initial rate F_0 is known and the other parameters $\{v_0, \beta, \gamma, \rho\}$ serve for calibration. The initial stochastic volatility v_0 is also known as the parameter α.

A natural choice of the rate behavior at zero is an *absorbing boundary condition*, which guarantees the martingale property of the rate. This also imposes a *finite* probability of the rate being at zero.

Like earlier (Chap. 3), we transform the SABR rate process F_t (4.1 and 4.2) into a stochastic volatility Bessel process (BES SV) Q_t,

$$Q_t = \frac{F_t^{1-\beta}}{1-\beta}, \tag{4.3}$$

which satisfies SDE

$$dQ_t = \left(\nu + \frac{1}{2}\right) Q_t^{-1}\, v_t^2\, dt + v_t dW_t^1, \tag{4.4}$$

$$dv_t = \gamma\, v_t\, dW_t^2, \tag{4.5}$$

with the *Bessel index* $\nu = -\frac{1}{2(1-\beta)}$.

We will denote marginal probability density functions (PDF's) of the processes F_t and Q_t as $p(t,f) = \mathbb{E}[\delta(F_t - f)]$ and $p(t,q) = \mathbb{E}[\delta(Q_t - q)]$ respectively. Here we use $\delta(x)$ to denote the Dirac delta-function. The SABR and Bessel SV PDF's are related via the probability elements equality, $p(t,f)\, df = p(t,q)\, dq$, which gives the following transformation rule:

$$p(t,f) = p(t,q)\, f^{-\beta}. \tag{4.6}$$

4.2 SABR Probability Density: Absorbing and Reflecting Solutions

The SABR density asymptotic properties are similar to these of the CEV model. In this section starting with the SDE (4.4), we address boundary conditions in terms of the PDF behavior at zero, identify them with absorption and reflection, and comment on the norm and moment conservation.

The BES SV process gives rise to the Forward Kolmogorov equation

$$p_t = -\left(\nu + \frac{1}{2}\right) v^2 \left(q^{-1} p\right)_q + \frac{1}{2} v^2 p_{qq} + \rho\gamma \left(v^2 p\right)_{qv} + \frac{1}{2}\gamma^2 \left(v^2 p\right)_{vv}, \quad (4.7)$$

which delivers a solution for the density $p(t, q, v) = E[\delta(q_t - q)\,\delta(v_t - v)]$ with the initial condition $p(0, q, v) = \delta(q_0 - q)\,\delta(v_0 - v)$. The solution is unique provided that certain boundary conditions are imposed at $q = 0$.

As in the pure Bessel case, we look for a solution at small q in the form

$$p(t, q, v) = q^\kappa\, \phi(t, q, v),$$

where function $\phi(t, q, f)$ is regular at $q = 0$. A balance of leading terms (of the order of $q^{\kappa-2}$) determines two possible characteristic exponents, $\kappa_1 = 1$ and $\kappa_2 = 2\nu + 1$, and thus gives rise to the following solutions

$$p^{(1)} = q\left(C_0 + C_1 q + O(q^2)\right), \quad (4.8)$$

$$p^{(2)} = q^{2\nu+1}\left(B_0 + B_1 q + O(q^2)\right). \quad (4.9)$$

Note that the second one may be realized only for $\nu > -1$ as follows from the integrability condition, $2\nu + 1 > -1$. Considering the next order leads to

$$\begin{aligned} v^2 C_1 &= -\frac{2\rho}{1 - 2\nu}\left(v^2 C_0\right)_v, \\ v^2 B_1 &= -2\rho\left(v^2 B_0\right)_v. \end{aligned} \quad (4.10)$$

For the zero correlation the first order coefficients cancel out.

Let us examine these asymptotics of being absorbing or reflecting. We notice that the marginal distribution of the stochastic volatility v_t is log-normal

$$p(t, v) = \int \mathrm{d}q\; p(t, q, v) = \frac{1}{\sqrt{2\pi t\gamma^2}}\, v^{-1}\, \mathrm{e}^{-\frac{1}{2}\frac{\left(\log v + \frac{1}{2}t\gamma^2\right)^2}{t\gamma^2}}. \quad (4.11)$$

This means that for any fixed q the PDF $p(t, q, v)$ goes to zero for $v \to 0$ with all its derivatives over v. This property permits us to understand the asymptotic behavior of the marginal distribution of q_t

$$p(t, q) = \int \mathrm{d}v\; p(t, q, v). \quad (4.12)$$

Indeed, integrating the Forward Kolmogorov equation over v, we obtain[1]

[1]We used the boundary properties of the density for $v \to 0$ to zero the two last terms.

$$\partial_t\, p(t,q) = \int \mathrm{d}\nu\, \nu^2 \left(-\left(\nu + \frac{1}{2}\right) \left(q^{-1}\, p\right)_q + \frac{1}{2} p_{qq} \right). \tag{4.13}$$

A time dependence of the norm

$$n(t) = \int \mathrm{d}q\, \mathrm{d}\nu\, p(t,q,\nu)$$

is established by the integration of the Eq. (4.13) over q and occurs to be dependent on the PDF behavior at the $q = 0$ boundary,

$$\partial_t\, n(t) = \int \mathrm{d}\nu\, \nu^2 \left(\left(\nu + \frac{1}{2}\right) q^{-1}\, p - \frac{1}{2}\, p_q \right)_{q\to 0}. \tag{4.14}$$

For the solution (4.8), we get

$$\partial_t\, n^{(1)}(t) = \nu \int \mathrm{d}\nu\, \nu^2 C_0(t,\nu)$$

while for the solution (4.9) the factor to be integrated becomes

$$\left(\left(\nu + \frac{1}{2}\right) q^{-1}\, p - \frac{1}{2}\, p_q \right)_{q\to 0} = -\frac{1}{2} q^{2\nu+1} \left(B_1 + O(q)\right)$$

and does not necessarily turn into zero at $q \to 0$. Indeed, we have $-1 < 2\nu + 1 < 0$ in the interval $0 < \beta < \frac{1}{2}$. However, integrating over ν cancels the potentially singular term due to (4.10)

$$\int \mathrm{d}\nu\, \nu^2 B_1(t,\nu) = -2\rho \int \mathrm{d}\nu\, \left(\nu^2 B_0\right)_\nu = 0$$

and results in the norm conservation

$$\partial_t\, n^{(2)}(t) = 0.$$

Thus, the norm-conserving solution $p^{(2)}$ (4.9) is naturally identified as reflecting, and the solution $p^{(1)}$ (4.8) which reveals the norm defect (at negative ν) is identified as absorbing.

Though interested mainly in negative index ν ($\beta < 1$), we comment briefly on the case $\nu > 0$. Imagine that the total PDF is presented by some combination of the solutions $p^{(1)}$ and $p^{(2)}$. For small q, the leading term of $p^{(1)} \sim q$ dominates the leading term of $p^{(2)} \sim q^{2\nu+1}$ implying that

$$p = p^{(1)} + p^{(2)} \simeq C_0(t,\nu) q\,,$$
$$\partial_t\, n(t) = \partial_t\, n^{(1)}(t) + \partial_t\, n^{(2)}(t) = \nu \int \mathrm{d}\nu\, \nu^2 C_0(t,\nu).$$

These relations, however, are in conflict. On the one hand, C_0 must be positive to define positive PDF p. Both ν and C_0 being positive, the time derivative of the norm must also be positive, $\partial_t n(t) > 0$, which is probabilistically impossible. Indeed, if the initial distribution is normalized to one, $n(0) = 1$, the norm $n(t)$ has no room to grow further. Thus, the absorbing solution $p^{(1)}$ may be realized only at $\nu < 0$ ($\beta < 1$). In this case, the norm defect, $\partial_t n(t) < 0$, merely indicates that there is a finite probability for process q_t to be at zero which is natural for the absorbing solution, $P(q_t = 0) = 1 - n(t)$.

We conclude that for a positive index $\nu > 0$ ($\beta > 1$) there exists only reflecting solution $p^{(2)}$ while for $\nu < -1$ ($\frac{1}{2} < \beta < 1$) the only possible solution is the absorbing one $p^{(1)}$. In these intervals of the index ν the PDF p is completely determined by the inner dynamics of the random processes q_t and v_t with no freedom for an outside boundary condition at $q = 0$. In the interval $-1 < \nu < 0$ ($\beta < \frac{1}{2}$) both solutions are legitimate, and we have to impose a boundary condition at $q = 0$ to select the proper unique solution. As we have seen, a selection of the reflecting solution is associated with the requirement of the norm conservation. Below we prove that a selection of the absorbing solution (and ignoring the reflecting one) is related with the martingale property of the SABR process F_t. Indeed, in terms of the BES SV process we need to calculate the (-2ν)-th moment $m_{-2\nu}(t)$ as far as the rate process reads $F_t = q_t^{-2\nu}\,(-2\nu)^{2\nu}$.

Multiplying the Forward Kolmogorov equation (4.7) by $q^{-2\nu}$ and rearranging terms we obtain

$$q^{-2\nu}\,p_t = \frac{1}{2}\,v^2\,\left(q^{-2\nu+1}\,(p\,q^{-1})_q\right)_q + \left[\rho\gamma\,q^{-2\nu}\,\left(v^2\,p\right)_q + \frac{1}{2}\gamma^2\,q^{-2\nu}\,\left(v^2\,p\right)_v\right]_v . \tag{4.15}$$

Thus, the moment time-derivative

$$\partial_t\,m_{-2\nu}(t) = \int \mathrm{d}q\,\mathrm{d}v\,q^{-2\nu}\,p_t(t,q,v) = -\frac{1}{2}\int \mathrm{d}v\,v^2\,\left(q^{-2\nu+1}\,(p\,q^{-1})_q\right)_{q=0} \tag{4.16}$$

has the following form for each of the solutions $p^{(1)}$ and $p^{(2)}$

$$\partial_t\,m^{(1)}_{-2\nu}(t) = 0\,,$$
$$\partial_t\,m^{(2)}_{-2\nu}(t) = -\nu\int \mathrm{d}v\,v^2\,B_0(t,v).$$

This indicates that the SABR process is a global martingale for the absorbing solution and a strict local one for the reflecting solution.

Below we will consider the SABR model with *absorbing boundary* as the most coherent and standard one.

4.3 Heat Kernel Expansion

The heat kernel expansion (DeWitt [25]) is a small-time asymptotic approximation for parabolic partial differential equations. This is a regular recipe for general stochastic systems (see the review of Avramidi [15]) to obtain expansion of the probability density function (PDF) as a fundamental solution to the Kolmogorov equation. The PDF expansion for the SABR model was calculated in Henry-Labordere [43] and Paulot [67]. It can be written in our (q, v) variables (4.3) as[2]

$$p(q, v) = \frac{1}{\gamma v^2 \sqrt{1-\rho^2}} \frac{1}{2\pi t} \sqrt{\frac{s(q, v)}{\sinh s(q, v)}} \mathscr{P}(q, v)\, \mathrm{e}^{-\frac{s^2(q,v)}{2\gamma^2 t}} (1 + O(t)) \qquad (4.17)$$

where *geodesic distance* $s(q, v)$ in the leading term depends on the volatility of the processes Q_t and v_t, and parallel transport $\mathscr{P}(q, v)$ depends on the drifts.

We notice that the distance and the parallel transport do not depend on time. It is also worth mentioning that the heat kernel does not take into account the boundary conditions: for small time, the rate "cannot" approach the boundary. In principal, it is possible to obtain the higher orders in time,[3] however, the formulas are very complicated (see Paulot [67]).

The distance s is measured along the geodesic line between points (q_0, v_0) and (q, v) in the geometry on the hyperbolic surface defined by the SABR governing Eqs. (4.1 and 4.2),

$$\cosh s = \frac{[\gamma \delta q - \rho(v - v_0)]^2}{2(1-\rho^2) v\, v_0} + \frac{(v - v_0)^2}{2 v v_0} + 1\,, \qquad (4.18)$$

where

$$\delta q = \frac{K^{1-\beta} - F_0^{1-\beta}}{1-\beta}.$$

The parallel transport may be defined by its logarithm $\mathscr{A}$

$$\mathscr{P} = e^{-\mathscr{A}}.$$

Loosely speaking, the term $\mathscr{A}$ is an integral of the system drift over the most probable path connecting initial point (q_0, v_0) and final point (q, v). Detailed consideration permits one to express the parallel transport as

$$\mathscr{A} = -(\nu + 1/2) \log\left(\frac{q}{q_0}\right) + \mathscr{B} = \frac{1}{2} \log\left(\frac{F}{F_0}\right)^{\beta} + \mathscr{B}\,, \qquad (4.19)$$

[2]The element of probability is defined as $\mathrm{d}P = p(q, v)\,\mathrm{d}q\,\mathrm{d}v$.

[3]Presented here as $O(t)$.

where the non potential part $\mathscr{B}$ is given by the following integral

$$\mathscr{B} = -\tfrac{1}{2}\frac{\beta}{1-\beta}\frac{\rho}{(1-\rho^2)}\int_C \frac{-\rho \mathrm{d}q' + \mathrm{d}V'}{q'} \tag{4.20}$$

along the geodesic line C from original to current point.

It occurs that the geodesic line C is a semi-circle in coordinates (x, y) related to original (q, v) as

$$q = \sqrt{1-\rho^2}x + \rho y \text{ and } v = \gamma y\,.$$

Parametrizing the curve C via the angle φ' on the circle, allows to transform integral $\mathscr{B}$ (4.20) into the form that can be taken analytically,

$$\mathscr{B} = -\tfrac{1}{2}\frac{\beta}{1-\beta}\frac{\rho}{\sqrt{1-\rho^2}}\left(\varphi - \varphi_0 - \int_{\varphi_0+\alpha}^{\varphi+\alpha}\frac{\mathrm{d}\varphi'}{1+L\sin\varphi'}\right). \tag{4.21}$$

here φ and φ_0 are angles for the initial and final points,

$$L^{-1} = \frac{x_c\sqrt{1-\rho^2}}{R}, \tag{4.22}$$

and x_c and R (expressions omitted) are the center and radius of the geodesic circle.

When passing to the marginal q distribution, integration over volatility v is performed using the saddle point method (see also [43, 67]) implying that the main contribution is due to the shortest path realized by the optimal volatility

$$v_{\min}^2 = \gamma^2\delta q^2 + 2\rho\gamma\delta q\, v_0 + v_0^2 \tag{4.23}$$

with the *optimal geodesic distance* $s_{\min}$ being a function of the initial value of the rate F_0, the initial stochastic volatility value v_0 and the strike K,

$$s_{\min} = s(q, v_{\min}) = \left|\log\frac{v_{\min} + \rho v_0 + \gamma\delta q}{(1+\rho)v_0}\right|. \tag{4.24}$$

The *optimal parallel transport*, which also depends on the strike K, is given by (4.19) with $\mathscr{B} = \mathscr{B}_{\min}$,

$$\mathscr{A}_{\min} = \mathscr{A}(q, v_{\min}) = \frac{\beta}{2}\log(K/F_0) + \mathscr{B}_{\min}\,, \tag{4.25}$$

where

$$\mathscr{B}_{\min} = \mathscr{B}(q, v_{\min}) = -\tfrac{1}{2}\frac{\beta}{1-\beta}\frac{\rho}{\sqrt{1-\rho^2}}\left(\pi - \varphi_0 - \arccos\rho - I\right), \tag{4.26}$$

and integral I is given by

$$I = \begin{cases} \frac{2}{\sqrt{1-L^2}} \left(\arctan \frac{u_0+L}{\sqrt{1-L^2}} - \arctan \frac{L}{\sqrt{1-L^2}} \right) & \text{for } L < 1 \\ \frac{1}{\sqrt{L^2-1}} \log \left| \frac{u_0\left(L+\sqrt{L^2-1}\right)+1}{u_0\left(L-\sqrt{L^2-1}\right)+1} \right| & \text{for } L > 1 \end{cases} \tag{4.27}$$

and

$$u_0 = \frac{\delta q\, \gamma \rho + v_0 - v_{\min}}{\delta q\, \gamma \sqrt{1-\rho^2}}, \qquad L = \frac{v_{\min}}{q\, \gamma\, \sqrt{1-\rho^2}} \quad \text{and} \quad \varphi_0 = \arccos\left(-\frac{\delta q\, \gamma + v_0\, \rho}{v_{\min}} \right).$$

All pieces collected, the small-time expansion for a call option with strike K and maturity T looks like

$$\mathcal{O}(T, K) = \frac{T^{\frac{3}{2}}}{2\sqrt{2\pi}} \exp\left\{ -\frac{1}{2} \frac{s_{\min}^2}{T\gamma^2} - \log \frac{s_{\min}^2}{2\gamma^2} + \log\left(K^{\beta} \sqrt{v_0 v_{\min}}\right) - \mathscr{A}_{\min} \right\}. \tag{4.28}$$

The small time expansion works fine for small times, but for moderate and large ones it needs to be improved. We will describe it in the next section.

4.4 Improved Approximation for Non-zero Correlation

Here, we will present results for an option price approximation superior to the Heat-Kernel expansion based on the mapping technique [7].

Namely, we can exactly calculate the SABR option price for the zero correlation (see the previous Section) but a non-zero correlation SABR still requires an approximation. Using Hagan/Paulot approximations can lead to substantial errors, especially, for large maturities and far OTM/ITM strikes. That is why we apply the mimicking technique mentioned above to improve the Heat-Kernel based Hagan/Paulot approximation. The idea is to come up with another model having the same small time expansion for the option (mimicking model) and calculate the final result using the mimicking model. For example, Hagan used the Black-Scholes model or normal one for this. Paulot has proposed the CEV model as the mimicking model. We use the SABR model with zero correlation (SABR ZC) having similar characteristics and asymptotics to the initial SABR model.

Denote the zero correlation SABR parameters with a tilde. Then, we should match for the initial and mimicking models the small time expansions:

$$\frac{1}{2}\frac{\tilde{s}_{\min}^2}{T\tilde{\gamma}^2}+\log\frac{\tilde{s}_{\min}^2}{2\tilde{\gamma}^2}-\log\left(K^{\tilde{\beta}}\sqrt{\tilde{v}_0\tilde{v}_{\min}}\right)+\tilde{\mathscr{A}}_{\min}$$
$$=\frac{1}{2}\frac{s_{\min}^2}{T\gamma^2}+\log\frac{s_{\min}^2}{2\gamma^2}-\log\left(K^{\beta}\sqrt{v_0v_{\min}}\right)+\mathscr{A}_{\min}. \quad (4.29)$$

We fix the vol-of-vol $\tilde{\gamma}$ and the power $\tilde{\beta}$ in the mimicking model (the zero correlation SABR model) and look for time-expansion of the initial volatility

$$\tilde{v}_0=\tilde{v}_0^{(0)}+T\,\tilde{v}_0^{(1)}+\cdots \quad (4.30)$$

Denote the function appearing in the argument of the logarithm of the optimal geodesic distance (4.24) as

$$\phi=\frac{v_{\min}+\rho v_0+\gamma\,\delta q}{(1+\rho)v_0}, \quad (4.31)$$

i.e., $s_{\min}=|\log\phi|$. Similarly, for the zero correlation, $\tilde{s}_{\min}(\tilde{v}_0)=|\log\tilde{\phi}(\tilde{v}_0)|$, we have

$$\tilde{\phi}(\tilde{v}_0)=\sqrt{1+\left(\frac{\delta\tilde{q}\,\tilde{\gamma}}{\tilde{v}_0}\right)^2}+\frac{\delta\tilde{q}\,\tilde{\gamma}}{\tilde{v}_0}, \quad (4.32)$$

where

$$\delta\tilde{q}=\frac{K^{1-\tilde{\beta}}-F_0^{1-\tilde{\beta}}}{1-\tilde{\beta}}.$$

To organize the fit (4.29) in the main order, we should find the leading order of the mimicking-model initial volatility (4.30) such that the equation

$$\frac{1}{2}\frac{\tilde{s}_{\min}^2\left(\tilde{v}_0^{(0)}\right)}{T\tilde{\gamma}^2}=\frac{1}{2}\frac{s_{\min}^2}{T\gamma^2} \quad (4.33)$$

is satisfied. A solution of this equation follows from the fit condition $\tilde{\phi}\left(\tilde{v}_0^{(0)}\right)=\phi^{\frac{\tilde{\gamma}}{\gamma}}$

$$\tilde{v}_0^{(0)}=\frac{2\,\Phi\,\delta\tilde{q}\,\tilde{\gamma}}{\Phi^2-1}, \quad (4.34)$$

where we have denoted $\Phi=\phi^{\frac{\tilde{\gamma}}{\gamma}}$. To calculate its first correction, we notice that the mimicking-model parallel transport does not depend on the initial volatility (4.25 and 4.26) due to zero correlation

$$\tilde{\mathscr{A}}_{\min}=\frac{1}{2}\log(K/F_0)^{\tilde{\beta}}. \quad (4.35)$$

Then, expand in time the square of the optimal distance of the mimicking model

$$\frac{1}{2}\tilde{s}^2_{\min}\left(\tilde{v}_0^{(0)} + T\,\tilde{v}_0^{(1)}\right) = \frac{1}{2}\tilde{s}^2_{\min}\left(\tilde{v}_0^{(0)}\right) - \Omega\, T\,\frac{\tilde{v}_0^{(1)}}{\tilde{v}_0^{(0)}} + \cdots \tag{4.36}$$

where the derivative coefficient reads

$$\Omega = \frac{\Phi^2 - 1}{\Phi^2 + 1}\log\Phi. \tag{4.37}$$

Substituting this into (4.29), we obtain a fit equation for the free terms in time

$$-\frac{\Omega\,\tilde{v}_0^{(1)}}{\tilde{\gamma}^2\,\tilde{v}_0^{(0)}} - \log\left(K^{\tilde{\beta}}\sqrt{\tilde{v}_0^{(0)}\tilde{v}_{\min}^{(0)}}\right) + \tilde{\mathscr{A}}_{\min} = -\log\left(K^{\beta}\sqrt{v_0 v_{\min}}\right) + \mathscr{A}_{\min}\,, \tag{4.38}$$

where the optimal volatility can be further simplified

$$\tilde{v}_{\min}^{(0)} = \sqrt{\delta\tilde{q}^2\,\tilde{\gamma}^2 + \tilde{v}_0^{(0)\,2}} = \tilde{v}_0^{(0)}\,\frac{\Phi^2 + 1}{2\,\Phi}. \tag{4.39}$$

This immediately gives us the correction to the initial volatility

$$\frac{\tilde{v}_0^{(1)}}{\tilde{v}_0^{(0)}} = \tilde{\gamma}^2\,\frac{\log\left(K^{\beta}\sqrt{v_0 v_{\min}}\right) - \log\left(K^{\tilde{\beta}}\sqrt{\tilde{v}_0^{(0)}\tilde{v}_{\min}^{(0)}}\right) + \tilde{\mathscr{A}}_{\min} - \mathscr{A}_{\min}}{\Omega}. \tag{4.40}$$

The effective zero correlation initial volatility depends on a choice of the fixed parameters $\tilde{\beta}$ and $\tilde{\gamma}$. A good choice based primarily on our numerical experiments reduces to

$$\tilde{\beta} = \beta, \tag{4.41}$$

$$\tilde{\gamma}^2 = \gamma^2 - \frac{3}{2}\left\{\gamma^2\rho^2 + \sigma_{BS}(F_0)\,\gamma\rho\,(1-\beta)\right\}, \tag{4.42}$$

where effective BS ATM implied volatility $\sigma_{BS}(F_0) = V_0\,F_0^{\beta-1}$. The intuition behind our choice is the following: The same power β helps with asymptotics for small strikes. The vol-of-vol $\tilde{\gamma}$ choice is inspired by a fit of the ATM implied volatility curvature.

The ATM case is a cumbersome but straightforward limit $K \to F_0$. The leading order ATM value reads[4]:

$$\tilde{v}_0^{(0)}\Big|_{K=F_0} = v_0.$$

[4]Here, we explicitly set $\tilde{\beta} = \beta$.

The first ATM correction can be also expressed in simple terms:

$$\left.\frac{\tilde{v}_0^{(1)}}{\tilde{v}_0^{(0)}}\right|_{K=F_0} = \frac{1}{12}\left(1 - \frac{\tilde{\gamma}^2}{\gamma^2} - \frac{3}{2}\rho^2\right)\gamma^2 + \frac{1}{4}\beta\rho\, v_0\gamma\, F_0^{\beta-1}.$$

4.5 Numerical Experiments

In this section, we numerically compare different approximation techniques for the SABR option price for large maturities and a wide strike coverage. The input data are summarized in the table below.

Rate initial value	F_0	1
SV initial value	v_0	0.25
Vol-of-vol	γ	0.3
Correlations	ρ	−0.2
Skews	β	0.6, 0.9
Maturities	T	10Y and 20Y

We present the second-moment underlying CMS calculations, as well as the Black-Scholes implied volatility for European options $\mathscr{C}(T, K) = \mathbb{E}[(F_T - K)^+]$.

CMS convexity adjustments depend on the second moment of the rate process, which can be evaluated by the usual static replication formula by Hagan [40]

$$\mathbb{E}[F_T^2] = 2\int_0^\infty dK\,\mathbb{E}[(F_T - K)^+].$$

For the SABR ZC Map option approximation, one can use this formula directly for the second moment calculations without any heuristic tricks (e.g., strike domain limitations, tail replacements, etc.). The tiny negativity of certain density approximations for the SABR ZC Map does not influence the quality of the CMS calculations. Note that, for close-to-zero correlations and large skews, the big-strike tail is very fat, which produces a very slow convergence of the static replication integral.

In our numerical experiments, we compare the following methods:

- Monte Carlo simulation (MC)
- The Hagan et al. [39] form of the implied volatility expansion (regular leading order, ATM first correction)
- The Paulot [67]/Henry-Labordere [43] form of the implied volatility expansion (regular leading order and the first correction)
- Map to the zero correlation SABR model (SABR ZC Map) (regular leading order and the first correction).

Table 4.1 Centered second moment and its errors for different methods

Maturity	Values		Error w.r.t. MC	
	10Y	20Y	10Y	20Y
MC	0.7637	1.028		
Paulot	0.8162	1.255	0.0525	0.227
Hagan	0.8263	1.733	0.0626	0.705
SABR zc Map	0.7817	1.065	0.018	0.037

In the Table 4.1, we summarize results for the centered second moment $\mathbb{E}[(F_T - F_0)^2]$ calculated for correlations $\rho = -0.5$ and skew $\beta = 0.6$.

In the tables below, we summarize results for the implied volatility, computed for different methods at various strikes, long maturity 20Y, and skews $\beta = 0.6$ (Table 4.2) and $\beta = 0.9$ (Table 4.3).

Table 4.2 Implied vol and its error for different methods, 20Y maturity, $\beta = 0.6$, $\rho = -0.2$

K	Value (%)				Error (bps)		
	MC	Paulot	Hagan	SABR ZC Map	Paulot	Hagan	SABR ZC Map
0.1	41.22	56.03	58.97	39.55	1481	1775	−167
0.2	35.58	46.2	48.86	34.29	1062	1328	−129
0.3	32.17	40.81	42.96	31.11	864	1079	−106
0.4	29.73	37.17	38.85	28.85	744	912	−88
0.5	27.86	34.49	35.75	27.13	663	789	−73
0.6	26.37	32.43	33.33	25.78	606	696	−59
0.7	25.16	30.8	31.39	24.69	564	623	−47
0.8	24.18	29.5	29.84	23.83	532	566	−35
0.9	23.38	28.45	28.59	23.13	507	521	−25
1	**22.73**	**27.61**	**27.61**	**22.59**	**488**	**488**	**−14**
1.1	22.22	26.93	26.84	22.17	471	462	−5
1.2	21.81	26.4	26.26	21.85	459	445	4
1.3	21.5	25.98	25.83	21.61	448	433	11
1.4	21.26	25.65	25.52	21.44	439	426	18
1.5	21.09	25.39	25.31	21.32	430	422	23
1.6	20.97	25.2	25.18	21.24	423	421	27
1.7	20.89	25.06	25.12	21.2	417	423	31
1.8	20.84	24.96	25.1	21.18	412	426	34
1.9	20.81	24.89	25.12	21.19	408	431	38
2	20.81	24.85	25.17	21.21	404	436	40

Table 4.3 Implied vol and its error for different methods, 20Y maturity, $\beta = 0.9$, $\rho = -0.2$

K	Value (%)				Error (bps)		
	MC	Paulot	Hagan	SABR ZC Map	Paulot	Hagan	SABR ZC Map
0.1	36.25	42.87	48.26	34.62	662	1201	−163
0.2	31.82	37.41	41.19	30.62	559	937	−120
0.3	29.19	34.27	37.05	28.25	508	786	−94
0.4	27.34	32.12	34.18	26.61	478	684	−73
0.5	25.96	30.54	32.03	25.39	458	607	−57
0.6	24.89	29.35	30.38	24.47	446	549	−42
0.7	24.06	28.44	29.11	23.78	438	505	−28
0.8	23.42	27.75	28.13	23.26	433	471	−16
0.9	22.93	27.23	27.39	22.88	430	446	−5
1	**22.56**	**26.85**	**26.85**	**22.62**	**429**	**429**	**6**
1.1	22.3	26.58	26.48	22.45	428	418	15
1.2	22.12	26.4	26.25	22.35	428	413	23
1.3	22.01	26.29	26.12	22.31	428	411	30
1.4	21.96	26.23	26.08	22.32	427	412	36
1.5	21.94	26.22	26.11	22.35	428	417	41
1.6	21.96	26.24	26.19	22.41	428	423	45
1.7	22.01	26.29	26.31	22.49	428	430	48
1.8	22.08	26.36	26.46	22.59	428	438	51
1.9	22.16	26.45	26.63	22.69	429	447	53
2	22.24	26.55	26.81	22.81	431	457	57

In all three tables, we observe an excellent approximation quality for the SABR ZC Map and insufficient approximation accuracy for Hagan or Paulot methods.

4.6 Conclusion

In this section we covered general asymptotic properties of the SABR model, studied its Heat-Kernel expansion and applied an expansion map for option prices approximation for non-zero correlation SABR (zero-correlation SABR exact price was studied in the previous section). We have checked the accuracy for the option pricing itself and the second moment underlying the CMS payment pricing.

Chapter 5
Extending SABR Model to Negative Rates

5.1 Introduction

At the time when the SABR model

$$dF_t = F_t^{\beta} \, v_t \, dW_t^1, \tag{5.1}$$

$$dv_t = \gamma \, v_t \, dW_t^2, \tag{5.2}$$

was introduced, positivity of the rates seemed like a reasonable and an attractive property. When rates are extremely low and even negative, it is important to extend the SABR model to negative rates. For example, on the graph below one can see a historical evolution of the Swiss Franc (CHF) interest rates (overnight and Libors of tenors 1M, 3M, and 6M). We see that negative rates reach in some cases −2%. Another important observation is that the rates "stick" to zero level for certain periods of time, suggesting their probability density functions having a singularity at zero (Fig. 5.1).

The simplest way to take into account negative rates is to shift the SABR process

$$dF_t = (F_t + s)^{\beta} \, v_t \, dW_t^1 \,,$$

where s is a deterministic positive shift. This moves the lower bound on F_t from 0 to $-s$.

One can either include the shift into calibration parameters ($v_0, \beta, \rho, \gamma, \mathbf{s}$) or fix it prior to calibration (e.g. to 2% in case of short rates of the Swiss Franc). Both ways have drawbacks.

Calibrating the shift does not really introduce a new degree of freedom—its influence on the skew is very similar to the power β, and may result in identification problem.

If we select the shift value *manually* and calibrate only the standard parameters (v_0, β, ρ, γ), there is always a danger that the rates can go lower than we anticipated, and we will need to change this parameter accordingly. This can result in jump in the other SABR parameters as the calibration response to such readjustment. As a consequence, we can get jumps in values/Greeks of the trades dependent on the

A. Antonov et al., *Modern SABR Analytics*,
SpringerBriefs in Quantitative Finance,
https://doi.org/10.1007/978-3-030-10656-0_5

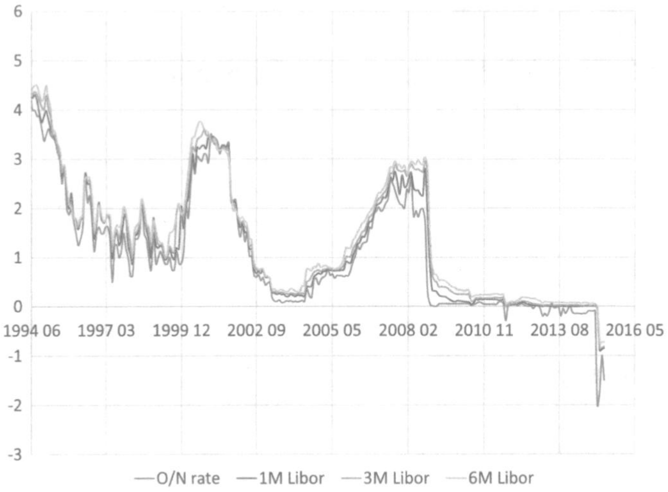

Fig. 5.1 Swiss Franc interest rates

swaption or cap volatilities. To cover for potential losses in such situations, traders are likely to be asked to reserve part of their P&L. Also, having the swaption prices being bounded from above (due to rate being bounded from below) can lead to situations when the shifted SABR cannot attain market prices. To sum this up, we need more natural and elegant solution permitting the negative rates.

For $\beta = 0$, the normal SABR model, $\mathrm{d}F_t = v_t \,\mathrm{d}W_t^1$, allows the rates to become negative when a free boundary condition is enforced. Below we come up with a generalization of this model

$$\mathrm{d}F_t = |F_t|^\beta \, v_t \,\mathrm{d}W_t^1$$

with $0 \leq \beta < \frac{1}{2}$, and a *free* boundary. As we will see, such model allows for negative rates and contains a certain "stickiness" at zero. Moreover, it is norm conserving and martingale.

In what follows, we consider only the $F_0 > 0$ case (unless explicitly stated otherwise). When $F_0 < 0$, we note that $\tilde{F}_t = -F_t$ satisfies the SABR SDE with parameters $(-F_0, v_0, \beta, -\rho, \gamma)$, and the *time-value* of a European option (call or put) on F_t struck at K equals that of an option on $\tilde{F}_t$ struck at $-K$. We do not distinguish between call and put *time-values* because, for norm conserving and martingale processes, they coincide, $\mathbb{E}[(F_T - K)^+] - (F_0 - K)^+ = \mathbb{E}[(K - F_T)^+] - (K - F_0)^+$.

To get intuition about the free boundary, we start with the CEV example $\mathrm{d}F_t = |F_t|^\beta \,\mathrm{d}W_t$ and study the PDF and the option prices. Then we switch to the SABR model with free boundary condition and present an exact solution for the zero-correlation case. For the general case, we show an accurate approximation for European options prices. We demonstrate that the exact formula as well as its approximation can be presented in terms of 1D integral over elementary functions making it attractive for a fast calibration.[1] We finish with simulation schemes and numerical results.

5.2 CEV Process

To aid with intuition, we consider the CEV model $\mathrm{d}F_t = F_t^\beta \,\mathrm{d}W_t$ with $0 \le \beta < 1$. The forward Kolmogorov (FK) equation on the density $p(t, f)$

$$p_t - \frac{1}{2}\left(f^{2\beta}\, p\right)_{ff} = 0.$$

has two types of solutions depending on the boundary conditions; fixing the PDE (or SDE) alone is not sufficient to uniquely define the solution. One can show (e.g. [12]) that there are two distinct solutions with asymptotics $p_A \sim f^{1-2\beta}$ and $p_R \sim f^{-2\beta}$. We call the first solution "absorbing" and the second one "reflecting". The latter exists only for $\beta < \frac{1}{2}$; otherwise, the norm around zero diverges.

The asymptotics are closely related to conservation laws, which can be obtained by integrating the FK equation by parts with some payoffs $h(f)$. Consider first the norm case of $h(f) = 1$. It is easy to see the asymptotics of the absorbing solution leads to non-conservation of the norm, while the reflecting solution conserves the norm. For the first moment conservation, we take $h(f) = f$ and deduce that the asymptotics of the reflecting solution leads to non-conservation of the first moment (i.e., non-martingality), while the absorbing solution is a martingale.

The explicit PDFs for the CEV process as well as option values are discussed in Chap. 2 (see also [10, 52]).

Below we use expression (2.59) for option prices for absorbing/reflecting solutions via a 1D integrals (see also [11, 12])

$$\mathcal{O}_{A/R}(T, K) = \frac{\sqrt{K F_0}}{\pi}\left(\int_0^\pi \mathrm{d}\phi \frac{\sin(|\nu|\phi)\sin\phi}{b - \cos\phi} \mathrm{e}^{-\frac{\bar{q}(b-\cos\phi)}{T}} + \sin(|\nu|\pi)\int_0^\infty \mathrm{d}\psi \frac{\mathrm{e}^{\mp|\nu|\psi}\sinh\psi}{b + \cosh\psi} e^{-\frac{\bar{q}(b+\cosh\psi)}{T}}\right) \tag{5.3}$$

for index $\nu = -\frac{1}{2(1-\beta)}$ and parameters

$$\bar{q} = q_0\, q_K, \quad b = \frac{q_0^2 + q_K^2}{2\, q_0\, q_K}, \quad q_0 = \frac{F_0^{1-\beta}}{1-\beta} \quad \text{and} \quad q_K = \frac{K^{1-\beta}}{1-\beta}.$$

[1]Note that the SABR approximation [39] based on the Heat-Kernel expansion cannot be applied to the free SABR because it does *not* take into account the boundary conditions.

Now consider an extension of the CEV model to the entire real line by modifying the SDE as follows

$$dF_t = |F_t|^\beta \, dW_t \tag{5.4}$$

for $0 \le \beta < \frac{1}{2}$. The corresponding forward Kolmogorov (FK) equation is

$$\partial_t \, p(t, f) = \frac{1}{2} \left(|f|^{2\beta} \, p(t, f) \right)_{ff} . \tag{5.5}$$

A *norm conserving* and *martingale* solution that satisfies the FK equation with the initial condition $p(0, f) = \delta(f - F_0)$ can be constructed from the reflecting and absorbing solutions as

$$p(t, f) = \frac{1}{2} \left(p_R(t, |f|) + \text{sign}(f) \, p_A(t, |f|) \right), \tag{5.6}$$

To see why Eq. (5.6) should hold we can employ a purely probabilistic argument. For any $f > 0$, a reflecting path ending at f is equivalent to a free-boundary path ending at f or $-f$, and vice versa. For the absorbing case, we apply the reflection principal—take the probability of path ending at f and subtract the probability of path ending at f and touching zero, which is equivalent to a path ending at $-f$. Hence, we can write the linear system

$$\begin{aligned} p_R(t, f) &= p(t, f) + p(t, -f), \\ p_A(t, f) &= p(t, f) - p(t, -f), \end{aligned}$$

which yields the expression for the free boundary density upon solving.

The solutions for typical parameters are shown in Fig. 5.2.

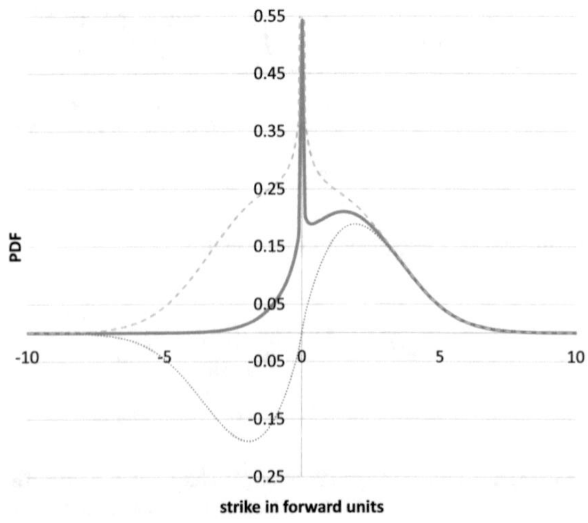

Fig. 5.2 The blue solid line represents the free PDF, the red dotted line depicts the absorbing density expression sign(f) $p_A(t, |f|)$ while the green dashed line gives the symmetrized reflecting solution

Taking a limit of $f f_0 \to 0$ in the Bessel functions underlying the absorbing and reflecting densities [12] we obtain the leading behavior of the Free CEV density

$$p(t, f, f_0) \underset{f f_0 \to 0}{=} |f|^{-2\beta} \left(C_1 + C_2 \, |f \, f_0|^{2(1-\beta)}\right) + C_3 \, \mathrm{sign}(f \, f_0) \, |f_0|^{\frac{1}{2}} \, |f|^{1-2\beta}.$$

We observe that for small f_0 the density become *symmetric* as function of f (the anti-symmetric absorbing part is attenuated due to small f_0) which leads to zero skew of the normal implied volatility.

Note also that at zero the PDF diverges as $p(t, f) \sim |f|^{-2\beta}$ (the asymptotic is inherited from the reflecting solution). The observed singularity is quite natural: one can observe a "sticky" behavior of the rates near zero (see Fig. 5.1 for CHF rate).

A call option payoff $h(f) = (f - K)^+$ leads to the option time value of

$$\begin{aligned} \mathcal{O}_F^{CEV}(T, K) &= \frac{1}{2}|K|^{2\beta} \int_0^T \mathrm{d}t \; p(t, K) \\ &= \frac{1}{2}|K|^{2\beta} \int_0^T \mathrm{d}t \; \frac{1}{2} \left(p_R(t, |K|) + \mathrm{sign}(K) \; p_A(t, |K|)\right) && (5.7) \\ &= \frac{1}{2} \left(\mathcal{O}_R(T, |K|) + \mathrm{sign}(K) \, \mathcal{O}_A(T, |K|)\right). && (5.8) \end{aligned}$$

Finally, we present the free CEV option time value using absorbing-reflecting solutions (5.3) and decomposition (5.8)

$$\mathcal{O}_F^{CEV}(\tau, K) = \frac{\sqrt{|K F_0|}}{\pi} \Bigg(\mathbf{1}_{K \geq 0} \int_0^\pi \mathrm{d}\phi \; \frac{\sin(|\nu| \, \phi) \sin \phi}{b - \cos \phi} \, \mathrm{e}^{-\frac{\bar{q}(b - \cos\phi)}{\tau}}$$
$$+ \sin(|\nu| \, \pi) \int_0^\infty \mathrm{d}\psi \; \frac{\left(\mathbf{1}_{K \geq 0} \cosh(|\nu| \, \psi) + \mathbf{1}_{K < 0} \sinh(|\nu| \, \psi)\right) \sinh \psi}{b + \cosh \psi} \, \mathrm{e}^{-\frac{\bar{q}(b + \cosh\psi)}{\tau}} \Bigg), \tag{5.9}$$

where $\nu = -\frac{1}{2(1-\beta)}$ and

$$\bar{q} = \frac{|F_0 \, K|^{1-\beta}}{(1-\beta)^2} \quad \text{with} \quad b = \frac{|F_0|^{2(1-\beta)} + |K|^{2(1-\beta)}}{2 \, |F_0 \, K|^{1-\beta}}.$$

We will use this formula to derive analytics for the SABR model in the section below. Note that we put the absolute value for F_0 for symmetry w.r.t. the strike: F_0 is considered to be positive according to the remark in the introduction.

Regarding a sensitive region of small strikes and/or small rates, we notice that the call option price (the full one, including the intrinsic value) is a smooth function of K and F_0 at zero. The thorough analysis reveals that main terms of expansion near zero are linear ones followed by terms of the order of $|K|^{2(1-\beta)}$ for small strikes and of $|F_0|^{2(1-\beta)}$ for small spots.

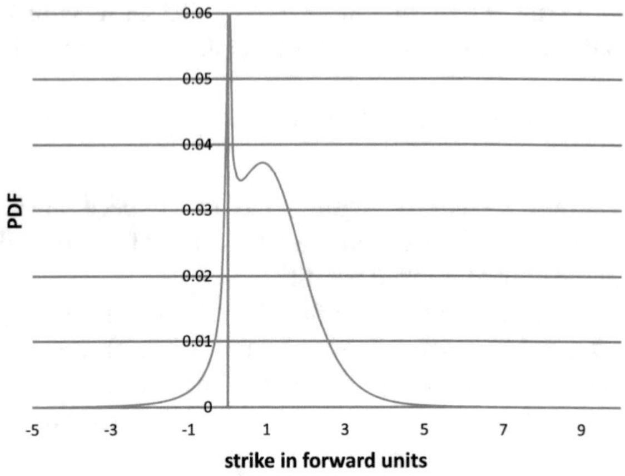

Fig. 5.3 SABR model PDF for $T = 3Y, \beta = 0.25$

5.3 SABR

Now, let us come back to the SABR process (5.1 and 5.2). The standard choice of the absorbing boundary will be generalized to the *free* boundary. Namely, we will consider the SDE

$$dF_t = |F_t|^\beta \, v_t \, dW_t^1$$

for $0 \le \beta < \frac{1}{2}$ (with the same process (5.2) for the stochastic volatility v_t). Such construction permits *negative* rates and "stickiness" at zero.

Looking forward, we plot the SABR density function, which is shown in Fig. 5.3 for the Input I parameters, as discussed in Sect. 5.6.2. We also observe the singularity which reflect a "sticky" behavior of the rates at zero (see Fig. 5.1).

5.3.1 *Zero-Correlation Case*

The zero-correlation free SABR model can be solved exactly. Indeed, the option price can be computed as

$$\mathcal{O}_F^{SABR}(T, K) = \mathbb{E}\left[\mathcal{O}_F^{CEV}(\tau_T, K)\right], \tag{5.10}$$

where $\mathcal{O}_F^{CEV}(\tau, K)$ is the free-boundary CEV option price (5.9) and the stochastic time $\tau_T = \int_0^T v_t^2 dt$ is the cumulative variance for the geometric Brownian motion v_t (5.2). Averaging over stochastic time τ_T results in the moment generating function (MGF) $\mathbb{E}\left[\exp\left(-\frac{\lambda}{\tau_T}\right)\right]$ which has been derived in [11] and discussed in Chap. 3 (3.108), recall that

$$\mathbb{E}\left[\exp\left(-\frac{\lambda}{\tau_T}\right)\right] = \frac{G(\gamma^2 T, s)}{\cosh s}, \quad \text{where} \quad s = \sinh^{-1}\left(\frac{\sqrt{2\lambda}\,\gamma}{v_0}\right). \tag{5.11}$$

Function $G(t, s)$, introduced in [10], represents the cumulative probability function on the hyperbolic plane H^2 and is derived in Chap. 3 (see (3.39)).

Thus, the exact option price for the zero correlation case can be presented as

$$\mathcal{O}_F^{SABR}(T, K) = \frac{1}{\pi}\sqrt{|K F_0|}\,\left\{\mathbf{1}_{K\geq 0} A_1 + \sin(|\nu|\pi)\, A_2\right\} \tag{5.12}$$

with integrals

$$A_1 = \int_0^\pi \mathrm{d}\phi \frac{\sin\phi \sin(|\nu|\phi)}{b - \cos\phi} \frac{G(\gamma^2 T, s(\phi))}{\cosh s(\phi)}, \tag{5.13}$$

$$A_2 = \int_0^\infty \mathrm{d}\psi \frac{\sinh\psi\,\left(\mathbf{1}_{K\geq 0}\cosh(|\nu|\psi) + \mathbf{1}_{K<0}\sinh(|\nu|\psi)\right)}{b + \cosh\psi} \frac{G(\gamma^2 T, s(\psi))}{\cosh s(\psi)}.$$

Here s has the following parametrization with respect to ϕ and ψ:

$$\sinh s(\phi) = \gamma\, v_0^{-1}\,\sqrt{2\bar{q}(b - \cos\phi)} \quad \text{and} \quad \sinh s(\psi) = \gamma\, v_0^{-1}\,\sqrt{2\bar{q}(b + \cosh\psi)},$$

where $\bar{q}$ and b are the same as in the CEV free-boundary option.

5.3.2 General Correlation Case

As in [10] and in Chap. 4, we approximate the general correlation option price by using the zero correlation one $\mathrm{d}\tilde{F}_t = |\tilde{F}|_t^{\tilde{\beta}}\,\tilde{v}_t\,\mathrm{d}\tilde{W}_1$ and $\mathrm{d}\tilde{v}_t = \tilde{\gamma}\,\tilde{v}_t\,\mathrm{d}\tilde{W}_2$, with $\mathbb{E}[\mathrm{d}\tilde{W}_1\,\mathrm{d}\tilde{W}_2] = 0$, i.e.,

$$\mathbb{E}\left[(F_t - K)^+\right] \simeq \mathbb{E}\left[(\tilde{F}_t - K)^+\right].$$

For the *free* boundary, we reuse the same effective coefficients of the zero-correlation SABR as in [10] for the *absorbing* boundary, with the only difference that we have to floor our strike, since all the effective parameters formulas based on the heat-kernel expansion work only for positive strikes. In our experiments, we used[2] $K_{\mathtt{eff}} = \max(K, 0.1\,F_0)$. The initial value of the rate F_0 is considered to be positive (see remark in the introduction for negative F_0).

[2]To avoid potential problems related with a non-smooth behavior around $F_0 = 10\,K$ we suggest $\max(K, 0.1F) \approx 0.1F + \frac{1}{2}\left(K - 0.1F + \sqrt{(K - 0.1F)^2 + \varepsilon^2}\right)$ for small parameter ε around 1bp.

Being a *real* process, the free SABR is naturally arbitrage-free. On the other hand, its approximation described above, strictly speaking, is not (except the case of the zero correlation when it becomes exact). However, given a high approximation accuracy, we can call the resulting analytical formula *quasi*-arbitrage-free.

5.3.3 Limiting Cases and Asymptotics

Below we briefly address behavior of the Free SABR call option $\mathscr{C}_F^{SABR}(T, K)$ for sensitive limiting cases.

Similar to the CEV model, the Free SABR call price is a smooth function of the strike and forward: one can show that

$$\begin{aligned}\mathscr{C}_F^{SABR} &\underset{K\to 0}{=} C_1 + C_2\, K + C_3 |K|^{2(1-\beta)} + \cdots \\ \mathscr{C}_F^{SABR} &\underset{F_0\to 0}{=} C_1' + C_2'\, F_0 + C_3' |F_0|^{2(1-\beta)} + \cdots\end{aligned}$$

where coefficients C_i and C_i' depend on the model parameters. This expansion follows from the CEV option expansion, with ‘SABR’ coefficients obtained as expectations of ‘CEV’ ones over the random time τ. Expansion is justified since moments $\mathbb{E}\left[\tau^a\right]$ exist for both positive and negative degrees a.

It is seen that call option “delta” is a smooth function of F_0 with the following behavior around zero:

$$\partial \mathscr{C}_F^{SABR} / \partial F_0 \underset{F_0\to 0}{=} C_2' + C_3'\, 2(1-\beta)\,\mathrm{sign}(F_0) |F_0|^{1-2\beta} + \cdots$$

The option “Gamma” is smooth everywhere except zero. This weak (integrable) divergence around zero, $\partial^2 \mathscr{C}_F^{SABR} / \partial F_0^2 \sim |F_0|^{-2\beta}$, reflects the rate “stickiness” On the other hand, a standard way of Greeks calculations based of finite differences with 1–5 bps spacing produces a moderate, finite “Gamma” spike at zero.

We have mentioned that the CEV model for the zero spot case has a symmetric density function and, as consequence, has zero implied volatility skew as zero strike. However, for the SABR model itself the asymmetry is introduced by the correlation with the stochastic volatility. This means that, for small or zero spots, the model can control the normal implied volatility skew around zero strikes by means of the correlation.

The case $\beta = 0$ is clearly regular. More interesting case of $\beta = \frac{1}{2}$ corresponds to a “merge” of absorbing and reflecting solutions to the absorbing one (for $\beta > \frac{1}{2}$ only absorbing solution exists). Thus, by construction (5.6), the free solution coincides the the absorbing one for $\beta \geq \frac{1}{2}$.

5.4 Mixture SABR

Instead of mapping a nonzero correlation SABR into a zero-correlation one, we can *define* our model as a mixture of a zero-correlation SABR and a normal SABR. Assume the forward rate F_t can be written as

$$F_t = \chi F_t^{(1)} + (1-\chi) F_t^{(2)},$$

where[3]

- $F_t^{(1)}$ follows a *zero-correlation* Free SABR with parameters $(\alpha_1, \beta_1, 0, \gamma_1)$,
- $F_t^{(2)}$ follows a *normal* Free SABR with parameters $(\alpha_2, 0, \rho_2, \gamma_2)$,
- χ is a random variable taking value 1 with probability p and 0 with probability $1-p$ and independent of both SABR processes.

In other words, at origin, we choose one of two "branches": with probability p we go with the zero correlation Free SABR and with probability $1-p$ we use the Normal Free SABR, and stay with the chosen model forever. The option price for the Mixture model is thus a weighted sum:

$$\begin{aligned}\mathbb{E}\left[(F_T-K)^+\right] &= \mathbb{E}\left[\chi\left(F_T^{(1)}-K\right)^+\right] + \mathbb{E}\left[(1-\chi)\left(F_T^{(2)}-K\right)^+\right] \\ &= p\,\mathbb{E}\left[\left(F_T^{(1)}-K\right)^+\right] + (1-p)\,\mathbb{E}\left[\left(F_T^{(2)}-K\right)^+\right].\end{aligned}$$

A motivation for this model is that its both branches are SABR processes, this model is arbitrage-free, allows rates go negative, and has dynamics similar to the Free/Shifted SABR (see the end of Sect. 5.6).

Both component models have closed-form solutions for option values (5.12) and (3.91) implying that the mixture model has an analytical solution[4]

$$\begin{aligned}\mathcal{O}_M(T,\ K;\ \alpha_1, \beta_1, \gamma_1;\ \alpha_2, \rho_2, \gamma_2) &= p\,\mathcal{O}_F(T,\ K;\ \alpha_1, \beta_1, 0, \gamma_1) \\ &\quad + (1-p)\,\mathcal{O}_N(T,\ K;\ \alpha_2, \rho_2, \gamma_2)\,, \qquad (5.14)\end{aligned}$$

written as three 1D integrals containing the function $G(t, s)$ (3.82), which can be *very* efficiently approximated by a closed formula (see [10]) with errors under 1bp. This approximation is far superior to other ones in the SABR "business", e.g. the Free SABR approximation for general correlations [12] or the original Hagan one for the absorbing SABR [39].

[3] We will write here α instead of v_0.

[4] The normal model component can be obviously written via zero beta Free SABR form, $\mathcal{O}_N(T,\ K;\ \alpha_2, \rho_2, \gamma_2) = \mathcal{O}_F(T,\ K;\ \alpha_2, 0, \rho_2, \gamma_2)$.

In addition, the Mixture SABR gives extra degrees of freedom, which can be used for calibration to either a larger number of swaptions or to swaptions and CMS. We verify the latter with numerical experiments in Sect. 5.6.

Regarding the parameter choices for the Mixture SABR, it is always useful to keep the same ATM volatility for both models, leading to the following relationship between the α's:

$$\sigma_0 = \alpha_1 F_0^{\beta_1} = \alpha_2. \tag{5.15}$$

The other parameters can be independent for greater calibration freedom.

A useful parametrization of the probability as a function of a parameter s is

$$p(s) = \frac{\sigma_0\,\beta_1\,e^s}{\sigma_0\,\beta_1\,e^s + |\gamma_2\,\rho_2|}, \tag{5.16}$$

which guarantees the Mixture SABR degeneration to either zero-correlation or normal SABR when $\rho_2 = 0$ or $\beta_1 = 0$, respectively.

The probability p can be also used to control the singularity at zero. Fixing it at some small value reduces the singularity arising from the zero-correlation model. We recommend, however, using the parametrization (5.16) for calibration, because the singularity is indeed observed in the rates time series (Fig. 1 in [12]).

5.4.1 *Parameter Intuition*

Both branches of the model yield the same ATM volatility (because of the constraint (5.15)), so the Mixture SABR will have the same ATM volatility as well.

The zero-correlation model affects the smile skew through its power β_1 while the normal one works with the smile skew through its correlation. The mixture of these models dictates the above skew behaviors.

The vol-of-vols $\{\gamma_1, \gamma_2\}$ affect both the smile curvature and the edges. Recall that the large strike limit of the implied Black volatility for SABR is $\frac{\gamma}{1-\beta}$ (see [10]). Thus, the Mixture model, having one more γ-parameter than the Shifted and the Free SABR models, can *decouple* the smile curvature and the edges. This feature allows the Mixture SABR calibrate to both swaptions and CMS quotes. The same goal was achieved by the ZABR model [6] employing numerical solution for option prices.

5.4.2 *Reduced Parametrization*

There is also a *reduced parametrization* with constraints

$$\gamma_2 = \frac{\gamma_1}{1-\beta_1} \quad \text{and} \quad p = \frac{\sigma_0\,\beta_1}{\sigma_0\,\beta_1 + |\gamma_2\,\rho_2|}. \tag{5.17}$$

The first constraint ensures the same large strikes asymptotics for both models. The second corresponds to the probability parametrization (5.16) with $s = 0$, allowing recover the pure zero-correlation or normal cases, and keep the same skew around the ATM strikes. The reduced parametrization has the same number of parameters (and their similar meaning) as the Free SABR.

5.5 Comparing the Shifted, Free, and Mixture SABRs

In this section, we compare the Shifted, Free, and Mixture SABRs. As already discussed, the Shifted SABR has a shift parameter which is not calibrated but is selected manually, and can be invalidated if rates go lower than anticipated. In this situation, a new value of the shift has to be chosen, which will lead to jump in other parameters as well as values/Greeks of *all* volatility sensitive instruments.

The Free and Mixture SABR are free of this shortcoming—their parameters are either calibrated or set without any future incompatibility with the market.

The analytical formulas for the Shifted and the Free SABRs are approximations. Their quality is, in general, good, but it can deteriorate in some cases, especially in the wings of the distribution. Sometimes, an ad hoc adjustment is necessary. On the other hand, the Mixture SABR analytics are exact and free of such adjustments.

All three models can be calibrated to observed swaption quotes. However, the Shifted and the Free SABRs cannot calibrate to swaptions and CMS quotes—they lack parameters to control the behavior of the wings. The Mixture SABR has more flexibility and is suitable for such joint calibration. We address these points in the numerical experiments section below.

Finally, the Free and Mixture SABRs have a singularity at zero which corresponds to the "stickiness" of the rate process at zero observed in the historical rates data. Having said that, one can attenuate this singularity by decreasing the probability in the Mixture SABR if desired.

5.6 Numerical Experiments

5.6.1 Calibration to Real Data

We start with a real data example of 1Y15Y CHF swaption from 10-Feb-2015 with the forward of $F_0 = 0.56\%$. The swaption prices are quoted in terms of normal implied volatility (bps). We calibrate the free-boundary and shifted SABR respectively to this data using our analytical approximations. The output is presented below in Table 5.1 and in Fig. 5.4: the calibration errors are tiny for both models.

Calibrated $\alpha = v_0, \rho, \gamma$ and β are given in Table 5.2 (the value of the shift is 2%).

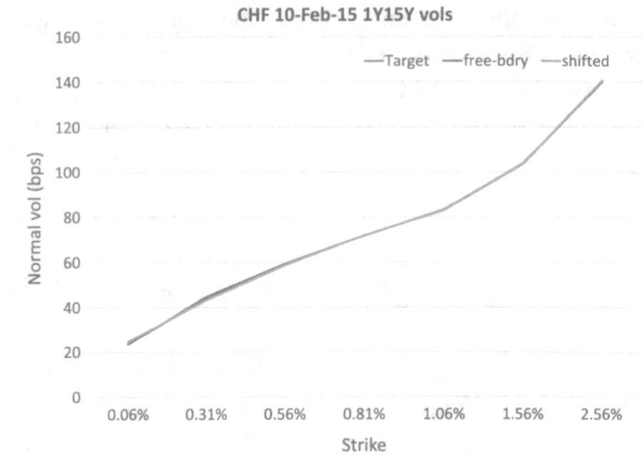

Fig. 5.4 Normal implied volatilities (bps)

Table 5.1 Normal implied volatilities (bps)

Strike (%)	Target	Free-bdry	Shifted
0.06	23.5	23.5	24.6
0.31	44.7	44.5	43.3
0.56	59.3	59.2	58.7
0.81	71.7	71.7	71.8
1.06	83.0	83.1	82.9
1.56	103.5	103.8	103.6
2.56	140.4	140.2	140.7

Table 5.2 Calibrated parameters

Param	Free SABR	Shifted SABR
α	0.051	0.011
β	0.417	0.167
ρ	0.990	0.999
γ	0.658	1.080

Note the extremely high values of the correlation ρ, and fairly high values of vol-of-vol γ. The reason for such high correlation is a very steep skew prevailing in the CHF market right now.

Now, we will study *accuracy* of the analytical approximation for the free SABR model. First, let us briefly address the Monte Carlo simulation scheme (see [12] for more details). Suppose that we have simulated the stochastic volatility for all time steps and paths v_t (this is trivial for the lognormal process). Our goal is to simulate $F_{t+\Delta t}$ given this info. The first thing to try is an Euler scheme without any boundary condition (free one) $F_{t+\Delta t} = F_t + |F_t|^\beta \, v_t \, \Delta W_1(t)$. One can check that the Euler scheme has an extremely slow convergence in both paths and time steps. Thus, we should come up with a more careful scheme based on numerical inversion of the CDF whose expression can be found in [12]. However, such procedure is very

Table 5.3 MC and ZC SABR simulations

K(%)	Analyt	Exact	Diff
0.06	24	24	−0.7
0.31	44	45	−0.7
0.56	59	60	−0.8
0.81	72	73	−0.8
1.06	83	84	−0.9
1.56	104	105	−1.0
2.56	140	142	−1.3

slow, and we use a regime switching scheme similar to [4] in order to accelerate the simulations. For out of boundary values, use the moment matching to approximate $F_{t+\Delta t}$ via the quadratic Gaussian step, while for near-boundary values, numerically invert the CDF.

In Table 5.3, we compare Monte Carlo simulations (Exact) described above and our analytical formula based on the map to the zero correlation SABR model (Analyt) for the calibrated parameters (see Table 5.1).

We observe an excellent approximation quality.

5.6.2 *Approximation Accuracy Analysis*

We provide the approximation accuracy analysis for two more inputs (somehow more "classical", e.g. with a negative correlation, see Table 5.4).

The implied volatility results are shown in Table 5.5 and plotted in Fig. 5.5.

We observe an excellent approximation quality for 3Y as well as for strikes $K > \frac{1}{2}F_0$ for 10Y. There is a slight degeneration for other strikes for 10Y. We see that the

Table 5.4 Setups for the free-boundary SABR model

Parameter	Symbol	Value for Input I	Value for Input II
Rate initial value	F_0	50 bps	1%
SV initial value	v_0	$0.6\,F_0^{1-\beta}$	$0.3\,F_0^{1-\beta}$
Vol-of-Vol	γ	0.3	0.3
Correlations	ρ	−0.3	−0.3
Skews	β	0.25	0.25
Maturities	T	3Y	10Y

Table 5.5 Differences in implied volatilities (in bps) between simulations (Exact) and analytics (Analyt). The bold line ($K = 1$) represents the ATM strike

K	Input I			Input II		
	Analyt	Exact	Diff	Analyt	Exact	Diff
−0.95	30.87	30.93	−0.06	40.05	40.86	−0.81
−0.8	29.83	29.95	−0.12	38.43	39.24	−0.81
−0.65	28.80	28.97	−0.17	36.80	37.60	−0.80
−0.5	27.79	27.99	−0.20	35.18	35.97	−0.78
−0.35	26.83	27.04	−0.21	33.59	34.33	−0.74
−0.2	25.95	26.15	−0.20	32.05	32.73	−0.68
−0.05	25.30	25.46	−0.16	30.67	31.25	−0.58
0.1	25.77	25.85	−0.08	30.20	30.63	−0.43
0.25	26.63	26.69	−0.06	30.19	30.51	−0.31
0.4	27.33	27.39	−0.06	30.14	30.41	−0.27
0.55	27.90	27.97	−0.06	30.06	30.31	−0.25
0.7	28.38	28.45	−0.07	30.00	30.22	−0.23
0.85	28.80	28.87	−0.07	29.98	30.18	−0.20
1	**29.18**	**29.25**	**−0.07**	**30.05**	**30.22**	**−0.17**
1.15	29.53	29.60	−0.07	30.24	30.36	−0.12
1.3	29.87	29.94	−0.07	30.56	30.63	−0.07
1.45	30.22	30.29	−0.06	31.03	31.04	−0.01
1.6	30.58	30.63	−0.06	31.63	31.58	0.04
1.75	30.95	30.99	−0.05	32.35	32.26	0.09
1.9	31.33	31.37	−0.04	33.17	33.04	0.13

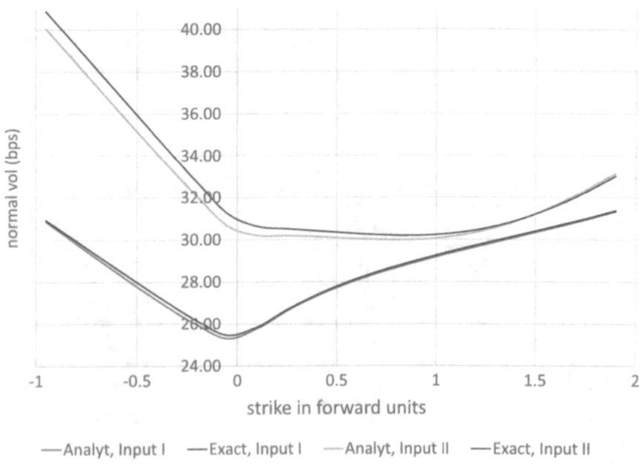

Fig. 5.5 Plot of implied volatility for Monte Carlo simulation (Exact) and our method (Analyt)

normal implied volatility possesses significant smiles with the bottom between zero and the ATM strike. In general, increasing the volatility-of-volatility and maturity moves the vertex of the smile to the ATM strike.

5.6.3 Mixture SABR Numerical Experiments

Next, we perform a number of numerical experiments where we compare our SABR models (Shifted, Free, and Mixture). The Mixture model is used either in the reduced form (5.17) or in the full one. We note that the ATM volatilities are always linked together (5.15).

We take real data from July 1, 2015 for Swiss Franc 1Y4Y swaptions market with negative forward rate of -11 bps. The input is presented in Table 5.6 in terms of normal implied volatility in pairs of strike & volatility, $\{K_i, \sigma_i\}$.

The first exercise is to calibrate our models to these swaptions. As shown in [13], the calibration accuracy is very good for all models. (The largest error is around 1 bp.)

A more challenging numerical experiment is a *joint* calibration to swaptions and a CMS payment—it allows us demonstrate the superiority of the Mixture SABR.

We recall that the CMS convexity adjustment depends on the variance of the rate process, which can be evaluated by the usual static replication formula [40]:

$$\mathbb{V}[F_T] \equiv \mathbb{E}[F_T^2] - F_0^2 = 2 \int_{-\infty}^{\infty} \mathrm{d}K \, \mathcal{O}(T, K). \tag{5.18}$$

As usual, $\mathcal{O}(T, K)$ is an option time value.

Unfortunately, CMS quotes are not always available in the market for a given swap rate. We analyze the ability of the Shifted, Free, and Mixture SABRs to reach different CMS prices. We quote the CMS payment in terms of its normal implied volatility σ_{CMS}, i.e.,

$$\mathbb{E}[F_T^2] - F_0^2 = T\sigma_{\text{CMS}}^2 \tag{5.19}$$

Table 5.6 1Y4Y swaption calibration input: strike versus implied normal volatility

Strike (%)	Vol (bps)
−0.3	56.42
−0.11	57.08
0.14	64.2
0.39	71.31
0.89	85.55
1.89	114.32

Determine the *minimal* volatility σ_{CMS} compatible with the input $\{K_i,\ \sigma_i\}_{i=1}^N$ from the Table 5.6. We start with constructing a curve of the implied volatility $\sigma(K)$ by using linear interpolation between the market nodes $\sigma(K_i) = \sigma_i$. Next, integrate the r.h.s. of (5.18) from the first quoted strike to the last one using the Bachelier formula for the option price:

$$\mathcal{O}(T, K) = \mathcal{O}_B(T, K, \sigma) \equiv \mathbb{E}[(F_0 - K + \sigma\sqrt{T}\,Z)^+] - (F_0 - K)^+, \qquad (5.20)$$

where Z is a standard Gaussian variable. The minimal CMS volatility, found by

$$T\tilde{\sigma}^2_{\text{CMS}} = 2\int_{K_1}^{K_N} \mathrm{d}K\ \mathcal{O}_B(T, K, \sigma(K)) \qquad (5.21)$$

with the input data in the Table 5.6, will be around sixty bps, $\tilde{\sigma}_{\text{CMS}} \simeq 60$ bps. Other interpolations, e.g., SABR ones, do not change the answer dramatically, so we can state that the CMS payment with a volatility smaller than 60 bps is *incompatible* with the swaptions input.

We examine a joint calibration to the set of swaptions and CMS volatilities from 50 to 100 bps, and analyze the results for all models: Shifted, Free, and Reduced Mixture and (full) Mixture.

We confirm our theoretical result that CMS volatilities smaller than the minimum value of ~60 bps cannot be reached by the calibration. (The exact calibration solution does not exist.) Moreover, for *attainable* CMS volatilities, the full Mixture model performs best. (The error is less than one bp.) The difference is due to a certain rigidity of the volatility curves of the Shifted, Free, and Reduced Mixture models, while the full Mixture model has enough degrees of freedom to maintain the calibrated swaption volatilities with almost independent movement of the wings for the CMS volatility calibration.

Table 5.7 shows the calibration errors for each model and indicates the clear accuracy winner: the Mixture model.

Below, we present the smiles of our models with the following correspondence of Series-CMS volatility:

Series number	1	2	3	4
Input CMS vol	70	80	90	100

We do not plot the calibrated volatilities for 50 and 60 bps of the CMS volatility as it is not compatible with the option volatility input. Instead, we concentrate on the larger CMS volatilities and examine how the model does the joint calibration by the smile deformation. Figures 5.6, 5.7, 5.8 and 5.9 show the calibrated volatilities for each of the SABR models (the underlying explicit numbers can be found in [13]).

We observe that only the full Mixture model is flexible enough to keep the calibrated option volatilities while adjusting the wings to address the changing CMS

Table 5.7 Calibration errors for the Shifted, Free, Reduced Mixture, and full Mixture SABR models (computed as *calibrated vol* – *input vol*, expressed in bps)

Input CMS vol (bps)		50	60	70	80	90	100
Model	Strike (%)						
Shifted	−0.3	−6.38	−4.61	−1.35	1.55	3.02	2.06
	−0.11	−0.98	0.42	1.27	1.67	1.27	0.01
	0.14	−0.58	0.61	0.01	−1.13	−2.45	−2.89
	0.39	−0.56	0.59	−0.33	−1.56	−2.44	−1.79
	0.89	−1.42	−0.11	−0.19	−0.07	0.72	3.06
	1.89	−5.70	−3.71	−0.41	3.27	7.36	12.15
Free	−0.3	−7.10	−4.88	−2.36	1.43	4.01	6.37
	−0.11	−1.50	0.10	1.70	2.94	2.86	3.56
	0.14	−0.82	0.37	−0.48	−1.97	−4.18	−4.97
	0.39	−0.55	0.53	0.75	−0.60	−2.34	−2.60
	0.89	−1.02	0.10	1.15	0.05	−0.78	−0.36
	1.89	−4.90	−3.35	−1.63	−1.00	1.52	3.62
Reduced Mixture	−0.3	−6.63	−4.45	−1.51	1.61	5.13	9.23
	−0.11	−1.01	0.19	1.04	1.02	0.72	0.08
	0.14	−0.42	0.28	−0.02	−1.72	−3.61	−6.34
	0.39	−0.31	0.32	−0.16	−0.62	−1.12	−2.34
	0.89	−1.14	−0.16	0.09	1.14	2.32	2.74
	1.89	−5.77	−3.47	−0.70	1.54	4.32	6.14
Mixture	−0.3	−6.69	−4.39	−1.21	−0.40	−0.22	−0.25
	−0.11	−0.98	0.45	0.99	0.99	0.90	1.13
	0.14	−0.36	0.37	−0.31	−0.70	−0.79	−0.56
	0.39	−0.26	0.09	−0.15	−0.15	−0.03	−0.22
	0.89	−1.12	−0.69	0.09	0.33	0.19	−0.42
	1.89	−5.81	−3.21	−0.61	−0.10	0.03	0.57

volatility. The other models do not have such flexibility and deform the option volatilities away from the target (Figs. 5.10, 5.11).

At the end we comment on dynamic properties of the Mixture SABR, i.e. the smile behavior under a displacement of the forward while the *other* model parameters being unchanged[5]. The SABR dynamics was one of its financial motivations, see [39]. For late 90's—early 00's the smile moved in the direction of the moving forward. However, it is not a market invariant: now the movement is different. Below we show the dynamics of the Free, Shifted and Mixture SABR's with the forward being shifted by 50 bps. We see that the results for all three models are similar—the smile follows the rate.

[5] As far as the SABR is mainly the strike interpolation (not a term-structure model) its process-level dynamic properties, i.e. conditional expectation $\mathbb{E}[(F_T - K)^+ \mid F_t = f]$, are irrelevant.

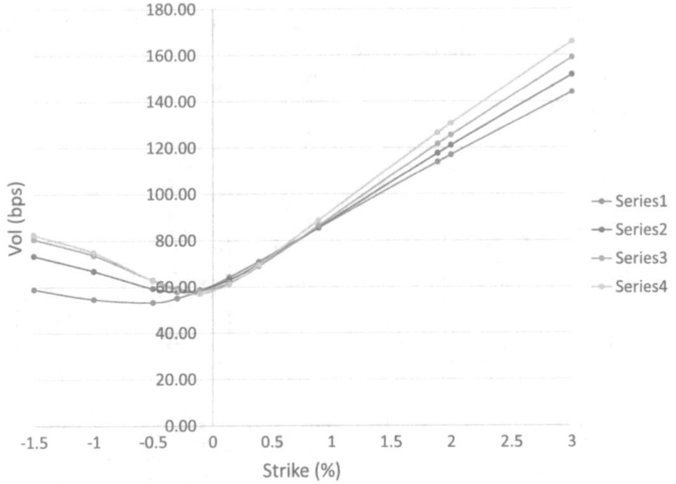

Fig. 5.6 Calibrated shifted SABR normal implied volatility

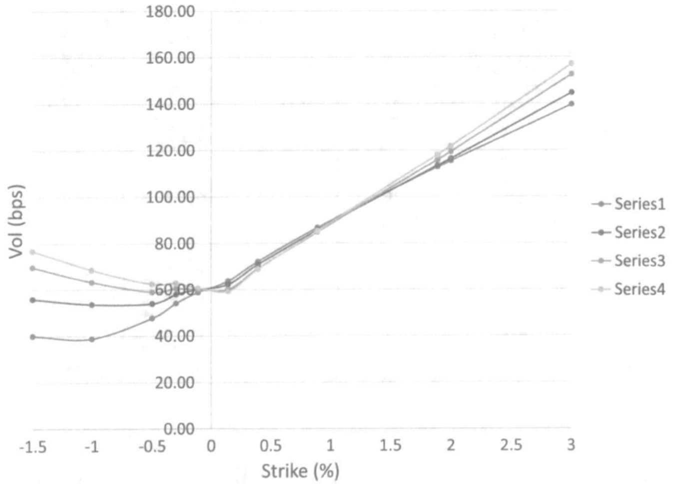

Fig. 5.7 Calibrated free SABR normal implied volatility

Introducing function $G(t, s)$ in Eq. (3.82), we have mentioned that, in practice, we use its 1D approximation (3.119) as derived in our earlier paper, [10], otherwise a 2D integration would render this model unfeasible. The numbers reported in this paper have been obtained with the approximation of $G(t, s)$. To show that this does not have much effect on the results, we present in Table 5.8 below normal volatilities for a wide range (± 3 standard deviations) of 10 year options for a reasonable set of parameters (Table 5.9) with both exact and approximate versions of $G(t, s)$. As one can see, the differences are well below 1 bp.

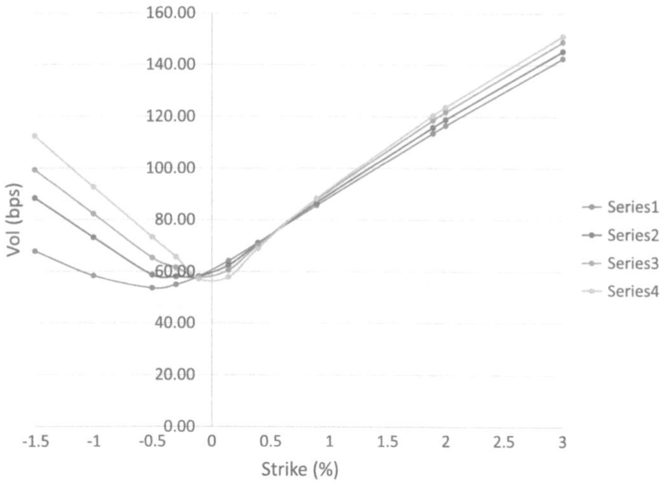

Fig. 5.8 . Calibrated reduced mixture SABR normal implied volatility

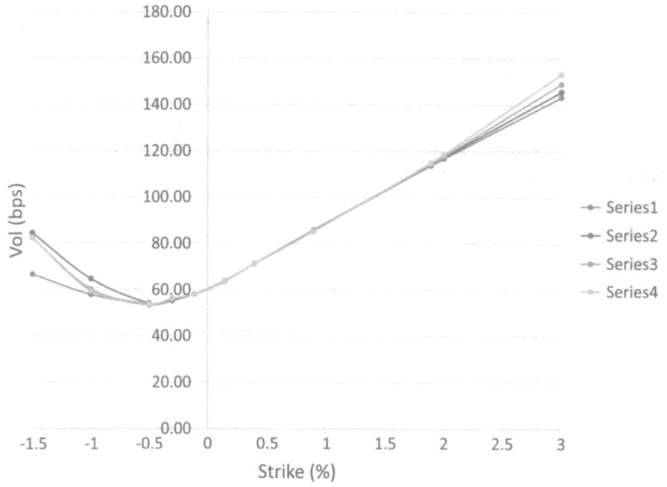

Fig. 5.9 Calibrated mixture SABR normal implied volatility

Since our method is based on a 1D integration, and the original Hagan's formula is written in closed form, our method is about 20–30 times slower, which is consistent with Gaussian quadrature requiring about 20–30 points for an accurate approximation of an integral. Adaptive quadrature methods, e.g. Gauss-Kronrod, may improve the speed of integration in some cases. If the exact function $G(t, s)$ is used, the integration becomes 2D making it slower than pricing using the approximate $G(t, s)$ by another factor of 20–30. Considering that we don't gain much precision, we recommend always using the approximation. Note that although theoretically using

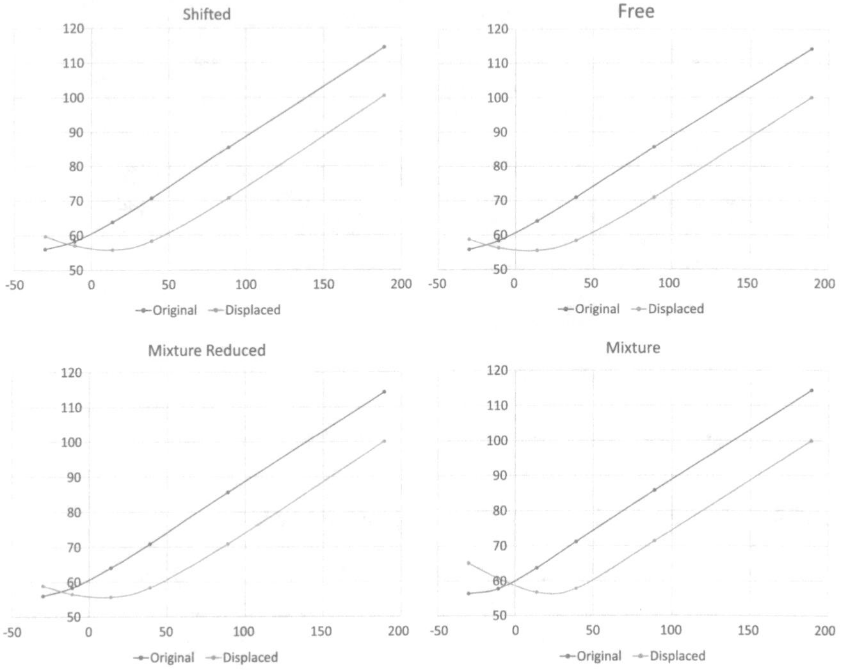

Fig. 5.10 Dynamics of the SABR normal implied volatility (bps) as function of strikes (bps): original forward versus displaced one

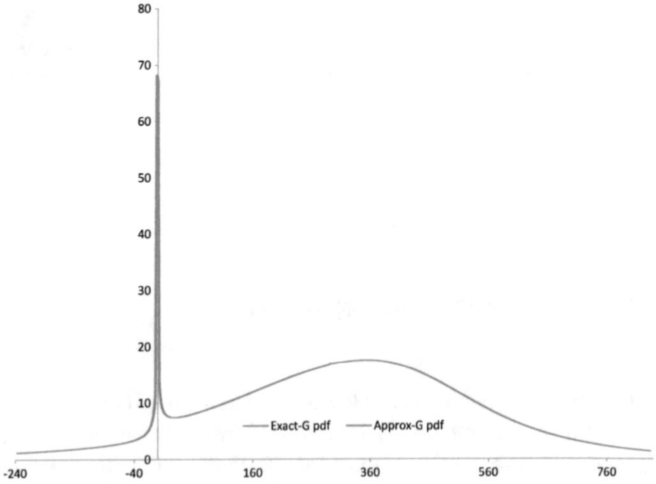

Fig. 5.11 Implied pdf of the 10Y swap rate with exact and approximate $G(t, s)$

Table 5.8 10Y results. All the results are in terms of normal volatilities. All values are in bps

Strike	Exact	Approx	Error
−506	148.77	148.79	0.01
−426	141.37	141.38	0.01
−345	133.82	133.84	0.01
−264	126.12	126.13	0.01
−184	118.25	118.26	0.01
−103	110.19	110.20	0.01
−23	102.01	102.02	0.01
58	96.77	96.78	0.01
139	93.03	93.05	0.01
219	89.89	89.91	0.01
300	87.64	87.65	0.01
381	86.67	86.69	0.01
461	87.48	87.49	0.01
542	90.31	90.32	0.01
623	94.87	94.88	0.01
703	100.62	100.64	0.01
784	107.09	107.11	0.01
864	113.97	113.98	0.01
945	121.07	121.08	0.01
1026	128.29	128.30	0.01
1106	135.55	135.56	0.01

Table 5.9 10Y Mixture SABR parameters

α_1	α_2	β	ρ	γ_1	γ_2	p
0.0346	0.0085	0.4	−0.5	0.3	0.3	0.35

an approximation of the function $G(t, s)$ we can no longer claim to guarantee the absence of arbitrage, in practice, the quality of our approximation is very good, and we have never encountered any arbitrage (negative probability density) in our tests. Below are the plots of the implied probability density obtained for the 10Y forward swap rate (with the same parameters as before) using the exact and approximate $G(t, s)$ implementations. The plots are practically indistinguishable.

Chapter 6
Conclusion

In this book, we have unified modern analytical material on the SABR model. Due to a small size of the book we have selected the shortest and most intuitive proofs of the underlying mathematical results. Moreover, we have presented simple insights for numerous non-trivial concepts in the stochastic processes and the approximation theory. Finally, we have presented numerous numerical results comparing analytics with simulation which can potentially serve as validation benchmarks.

We have started with the CEV model as a special case of the SABR and covered such important concepts as absorbing and reflecting boundaries, local and global marginality and a mall time asymptotic analysis. Then, we have switched to exactly solvable cases, such as zero correlation SABR, normal SABR with the free boundary and the log-normal SABR. A big chapter was devoted to the SABR approximation for small times including a comprehensive description of the Heat-Kernel, a simple derivation of the SABR option price for small maturities and its mapping approximation. We have finished the book with a Chapter describing the SABR generalization for negative rates including the free boundary SABR and the SABR Mixing model having an exact analytical option price.

As a future research in the SABR analytics, we would suggest the following directions:

- Adding a mean-reversion to the stochastic volatility
  Such modification will not change the model property on the small times but will bring the stochastic volatility process in line with empirical facts and make the forward distribution on the large dates less "aggressive". The main challenge here is the analytical approximation of the option price.
- Accelerate the CMS pricing
  As we have mentioned in the book, a CMS payment price can be statically replicated using the options. However, this procedure can be relatively slow for cali-

A. Antonov et al., *Modern SABR Analytics*,
SpringerBriefs in Quantitative Finance,
https://doi.org/10.1007/978-3-030-10656-0_6

bration purposes. Thus, a direct and fast calculation of the SABR second moment underlying the CMS payment can be useful and interesting analytical task.
- And finally, apply the fashionable Machine Learning techniques to the SABR universe!

The authors are indebted to Serguei Mechkov for discussions and numerical implementation help, as well as to their colleagues at Numerix, especially to Gregory Whitten and Serguei Issakov for supporting this work.

References

1. Abramowitz, M., Stegun, I. (eds.): Handbook of Mathematical Functions: With Formulas, Graphs, and Mathematical Tables. Dover Publications, New York (1970)
2. Andersen, L., Andreasen, J.: Volatility skews and extensions of the LIBOR market model. Appl. Math. Financ. **7**(1), 1–32 (2000)
3. Andersen, L., Brotherton-Ratcliffe, R.: Extended Libor market models with stochastic volatility. SSRN eLibrary (2001)
4. Andersen, L.: Efficient simulation of the Heston stochastic volatility model. SSRN paper (2007)
5. Andersen, L., Piterbarg, V.: Moment explosions in stochastic volatility models. Financ. Stoch. **11**(1), 29–50 (2007)
6. Andreasen, J., Huge, B: Expanded Forward Volatility. RISK Magazine, January (2013)
7. Antonov, A., Misirpashaev, T.: Projection on a quadratic model by asymptotic expansion with an application to LMM swaption. SSRN paper (2009)
8. Antonov, A., Spector, M.: Advanced analytics for the SABR model. In: WBS 7th Fixed Income Conference Presentation (2011)
9. Antonov, A., Spector, M.: Advanced analytics for the SABR model. SSRN (2012)
10. Antonov, A., Konikov, M., Spector, M.: SABR Spreads Its Wings. Risk Magazine, August (2013)
11. Antonov, A., Konikov, M., Rufino, D., Spector, M.: Exact solution to CEV model with uncorrelated stochastic volatility. SSRN paper (2014)
12. Antonov, A., Konikov, M., Spector, M.: The Free Boundary SABR: Natural Extension to Negative Rates. RISK Magazine, September (2015)
13. Antonov, A., Konikov, M., Spector, M.: Mixing the SABR for Negative Rates: Analytical Arbitrage-Free Solution. SSRN paper (2015). Also in RISK Magazine, April (2017)
14. Armstrong, J., Forde, M., Lorig, M., Zhang, H.: Small-time asymptotics under local-stochastic volatility with a jump-to-default: curvature and the heat kernel expansion. SIAM J. Financ. Math. **8**(1), 82–113 (2017)
15. Avramidi, I.: Heat Kernel Method and Its Applications. Birkhäuser, Basel (2015)
16. Balland, P., Tran, Q.: SABR goes Normal. Risk Magazine, May (2013)
17. Berestycki, H., Busca, J., Florent, I.: Computing the implied volatility in stochastic volatility models. Commun. Pure Appl. Math. **57**(10), 1352–1373 (2004)
18. Borodin, A., Salminen, P.: Handbook of Brownian Motion: Facts and Formulae. Probability and Its Applications, 2 rev edn. Birkhauser, Verlag AG (2002)
19. Carr, P., Schröder, M.: Bessel processes, the integral of geometric Brownian motion, and Asian options. Theory Probab. Appl. **48**, 400–425 (2004)
20. Cox, J.: The constant elasticity of variance option pricing model. J. Portf. Manag. **22**, 5–17 (1996)

A. Antonov et al., *Modern SABR Analytics*,
SpringerBriefs in Quantitative Finance,
https://doi.org/10.1007/978-3-030-10656-0

21. Cox, J., Ross, S.: The valuation of options for alternative stochastic processes. J. Financ. Econ. **7**, 229–263 (1976)
22. Davis, E.B.: Heat Kernels and Spectral Theory. Cambridge University Press (1989)
23. Davydov, D., Linetsky, V.: Pricing and hedging path-dependent options under the CEV process. Manag. Sci. **47**, 949–965 (2001)
24. De Marco, S., Hillairet, C., Jacquier, A.: Shapes of implied volatility with positive mass at zero. arXiv:1310.1020v4 [q-fin.PR] (2017)
25. DeWitt, B.: Dynamical Theory of Groups and Fields. Gordon and Breach (1965)
26. Döring, L., Horvath, B., Teichman, J.: Functional analytic (ir-)regularity properties of SABR-type processes. Int. J. Theor. Appl. Financ. **20**(3) (2017)
27. Doust, P.: No-arbitrage SABR. J. Comput. Financ. **15**(3), 3–31 (2012)
28. Dufresne, D.: The Integrated Square-Root Process. Centre for Actuarial Studies, Department of Economics, University of Melbourne (2001)
29. Dufresne, D.: Bessel processes and a functional of Brownian motion. In: Workshop on Mathematical Methods in Finance, Melbourne (2004)
30. Emanuel, D., MacBeth, D.: Further results on the constant elasticity of variance call option pricing model. J. Financ. Quant. Anal. **17**, 533–554 (1982)
31. Feller, W.: Two singular diffusion problems. Ann. Math. **54**(1), 173–182 (1951). Second Series
32. Geman, H., Shih, Y.F.: Modeling commodity prices under the CEV model. J. Altern. Invest. **11**(3), 65–84 (2009)
33. Goeing-Jaeschke, A., Yor, M.: A survey and some generalizations of bessel processes. Bernoulli **9**(2), 313–349 (2003)
34. Gradshteyn, I.S., Ryzhik, I.M., Jeffrey, A. (ed.): Tables of Integrals, Series, and Products, 6th edn. Academic Press, New York (2000)
35. Gruet, J.-C.: Semi-groupe du mouvement Brownien hyperbolique. Stoch. Stohastic Rep. **56**, 53–61 (1996)
36. Gulisashvili, A., Horvath, B., Jacquier, A.: Mass at zero in the uncorrelated SABR model and implied volatility asymptotics. Quant. Financ. **10**, 1753–1765 (2018)
37. Gulisashvili, A., Horvath, B., Jacquier, A.: On the probability of hitting the boundary for Brownian motions on the SABR plane. Electron. Commun. Probab. **21**(75), 1–13 (2016)
38. Gyöngy, I.: Mimicking the one-dimensional marginal distributions of processes having an Ito differential. Probab. Theory Relat. Fields **71**, 501–516 (1986)
39. Hagan, P., Kumar, D., Lesniewski, A., Woodward, D.: Managing Smile Risk, vol. 1, pp. 84–108. Wilmott Magazine, September (2002)
40. Hagan, P.: Convexity Conundrums: Pricing CMS Swaps, Caps, and Floors. Wilmott Magazine, pp. 38–44, March (2003)
41. Hagan, P., Kumar, D., Lesniewski, A., Woodward, D.: Arbitrage-Free SABR. Wilmott Magazine, pp. 60–75, January (2014)
42. Hagan, P., Lesniewski, A., Woodward, D.: Probability distribution in the SABR model of stochastic volatility. In: Working paper (2005). Also in Large Deviations and Asymptotic Methods in Finance, pp. 1–35. Springer (2015)
43. Henry-Labordère, P.: Analysis, Geometry, and Modeling in Finance: Advanced Methods in Option Pricing. Chapman and Hall/CRC (2008)
44. Heston, S.: A closed-form solution for options with stochastic volatility with applications to bond and currency options. Rev. Financ. Stud. **6**(2), 327–343 (1993)
45. Heston, S., Loewenstein, M., Willard, G.A.: Options and bubbles. Rev. Financ. Stud. **20**(2), 359–390 (2007)
46. Horvath, B.: Robust methods for the SABR model and related processes: analysis, asymptotics and numerics. Ph.D. thesis (2015). ETH Zürich
47. Horvath, B., Reichmann, O.: Dirichlet forms and finite element methods for the SABR model. SIAM J. Financ. Math. **9**(2) (2018)
48. Hull, J.: Options, Futures and Other Derivatives. Prentice Hall (2003)
49. Hull, J., White, A.: The pricing of options on assets with stochastic volatilities. J. Financ. **42**(2), 281–300 (1987)

50. Islah, O.: Solving SABR in exact form and unifying it with LIBOR market model. SSRN paper (2009)
51. Jarrow, R., Protter, P., Shimbo, K.: Asset price bubbles in incomplete market. Math. Financ. **20**(2), 145–185 (2010)
52. Jeanblanc, M., Yor, M., Chesney, M.: Mathematical Methods for Financial Markets. Springer (2009)
53. Jourdain, B.: Loss of martingality in asset price models with lognormal stochastic volatility. ENPC-CERMIX, Working paper (2004)
54. Korn, R., Tang, S.: Exact Analytical Solution for the Normal SABR Model. Wilmott Magazine, vol. 13, pp. 64–69 (2013)
55. Lesniewski, A.: Notes on the CEV model. In: Working paper. Ellington Management Group (2009)
56. Lewis, A.L.: Option Valuation under Stochastic Volatility I. Finance Press, Newport Beach (2000)
57. Lewis, A.L.: Option Valuation under Stochastic Volatility II. Finance Press, Newport Beach (2016)
58. Leitao Rodriguez, A., Grzelak, L., Oosterlee, C.: On a one time-step SABR simulation approach: application to European options. Appl. Math. Comput. **293**(January), 461–479 (2017)
59. Lindsay, A.E., Brecher, D.R.: Simulation of the CEV process and the local martingale property. Math. Comput. Simul. **82**, 868–878 (2012)
60. Lions, P.-L., Musiela, M.: Correlations and bounds for stochastic volatility models. Annales de l'Institut Henri Poincaré (C), Non Linear Analysis **24**(1), 1–16 (2007)
61. MacBeth, J., Merville, L.: Tests of the Black-Scholes and Cox call option valuation models. J. Financ. **35**(1980), 285301 (1980)
62. McKean, H.P.: An upper bound to the spectrum of Δ on a manifold of negative curvature. J. Diff. Geom. **4**, 359–366 (1970)
63. Matsumoto, H., Yor, M.: Exponential functionals of Brownian motion, I: probability laws at fixed time. Probab. Surv. **2**, 312–347 (2005)
64. Matsumoto, H., Yor, M.: Exponential functionals of Brownian motion, II: some related diffusion processes. Probab. Surv. **2**, 348–384 (2005)
65. Mercurio, F., Morini M.: Joining the SABR and Libor Models Together. Risk, pp. 80–85, March (2009)
66. Mercurio, F., Moreni, N.: Inflation modelling with SABR dynamics. Risk, pp. 106–111, June (2009)
67. Paulot, L.: Asymptotic Implied Volatility at the Second Order with Application to the SABR Model. SSRN paper (2009). Also in Large Deviations and Asymptotic Methods in Finance, pp. 37–69. Springer (2015)
68. Rebonato, R., McKay, K., White, R.: The SABR/LIBOR Market Model: Pricing, Calibration and Hedging for Complex Interest-Rate Derivatives. Wiley (2009)
69. Revuz, D., Yor, M.: Continuous Martingales and Brownian Motion. Springer (1999)
70. Schröder, M.: Computing the constant elasticity of variance option pricing formula. J. Financ. **44**(1), 211–219 (1989)
71. Sin, C.: Complications with stochastic volatility models. Adv. Appl. Probab. **30**, 256–268 (1998)
72. Skabelin, A.: Pricing of options on assets with level dependent stochastic volatility. In: Noise and Fluctuations in Econophysics and Finance, SPIE, vol. 5848, 110124 (2005). http://www.fma.org/Chicago/Papers/AlexanderSkabelinStochasticVolatility
73. Spector, M.: The Log-Normal SABR. Numerix working paper, New York (2011)
74. Watson, G.N.: A Treatise on the Theory of Bessel Functions, 2nd edn. Cambridge University Press (1944)
75. Yor, M.: On some exponential functionals of Brownian motion. Adv. Appl. Probab. **24**(3), 509–531 (1992)
76. Varadhan, S.R.S.: Diffusion processes in a small time interval. Commun. Pure Appl. Math. **20**, 659–685 (1967)

Printed in the United States
By Bookmasters